Köln, Oktober 202

Liebe Kim,

vielen Dank für die guten und inspirierenden Diskussionen zum Job und darüber hinaus, die wir in dem letzten Jahr hatten. Ich hoffe, dass wir diese noch einige Zeit fortführen können.

Auch wenn dieses Buch nicht gerade die spannendste Abendlektüre ist hoffe ich dennoch dass es dir ein paar neue Gedanken bringt.

Vielleicht auch in die Richtung selbst einmal ein Buch zu schreiben - ich würde mich sehr freuen deins in 3 Jahren lesen zu dürfen.

Jan

Sustainable Production, Life Cycle Engineering and Management

Series editors

Christoph Herrmann, Braunschweig, Germany
Sami Kara, Sydney, Australia

Modern production enables a high standard of living worldwide through products and services. Global responsibility requires a comprehensive integration of sustainable development fostered by new paradigms, innovative technologies, methods and tools as well as business models. Minimizing material and energy usage, adapting material and energy flows to better fit natural process capacities, and changing consumption behaviour are important aspects of future production. A life cycle perspective and an integrated economic, ecological and social evaluation are essential requirements in management and engineering. This series will focus on the issues and latest developments towards sustainability in production based on life cycle thinking.

More information about this series at http://www.springer.com/series/10615

Jan Beier

Simulation Approach Towards Energy Flexible Manufacturing Systems

Springer

Simulation Approach Towards Energy Flexible Manufacturing Systems

Von der Fakultät für Maschinenbau

der Technischen Universität Carolo-Wilhelmina zu Braunschweig

zur Erlangung der Würde

eines Doktor-Ingenieurs (Dr.-Ing.)

genehmigte Dissertation

von

Dipl. Wirtsch.-Ing. Jan Beier, MBA

aus Wolfenbüttel

eingereicht am: 31. März 2016

mündliche Prüfung am: 27. Mai 2016

Gutachter: Prof. Dr.-Ing. Christoph Herrmann

Prof. Sami Kara

Prof. Dr. Kuldip Singh Sangwan

2017

Jan Beier
Institute of Machine Tools and Production
Technology
Technische Universität Braunschweig
Braunschweig
Germany

ISSN 2194-0541 ISSN 2194-055X (electronic)
Sustainable Production, Life Cycle Engineering and Management
ISBN 978-3-319-46638-5 ISBN 978-3-319-46639-2 (eBook)
DOI 10.1007/978-3-319-46639-2

Library of Congress Control Number: 2016956388

Printed on acid-free paper

This Springer imprint is published by Springer Nature
The registered company is Springer International Publishing AG
The registered company address is: Gewerbestrasse 11, 6330 Cham, Switzerland

Foreword

Mitigating climate change has become a central part of sustainable development. Within this context, manufacturing plays a central role on the energy demand side since manufacturing activities require substantial material and energy inputs. Therefore, energy demand and resulting emissions from manufacturing need to be reduced to mitigate climate change. Besides reducing anthropogenic greenhouse gas emissions by increasing energy efficiency of processes, alternative processes are developed to replace high-emission processes. However, energy efficiency increase alone is insufficient to reduce emissions towards a desired target; hence energy generated from renewable sources needs to be supplied to factories to reduce emissions further. Most notably, a switch from non-renewable/fossil fuel based energy supply to renewable energy supply is required.

Traditionally, manufacturing companies regard energy, especially electricity supply, as a dispatchable resource. However, with the integration of renewable energy supply such as wind and solar resources, electricity supply becomes intermittent due to non-controllable factors such as weather conditions. One approach to cope with this intermittent energy supply is the involvement of customers by adjusting electricity demand to variable supply. Therefore, manufacturing companies can play a central role by adjusting their demand to energy availability, hence substantially contributing to the integration of renewable energy sources.

The concept presented by Jan Beier directly addresses this topic by allowing manufacturing organisations to plan their production with a more flexible energy demand and sustaining manufacturing activities under energy flexibility goals. The concept is centered around an improvement cycle, which establishes energy flexibility objectives and defines manufacturing system setups which are expected to increase energy flexibility. In order to enable and test hypotheses, a manufacturing system energy flexibility model and control is proposed. The model allows dynamic calculation of the manufacturing system behavior and thus determining relevant material and energy flows as well as related performance indicators under varying system setups.

Braunschweig, Germany
Sydney, Australia
July 2016

Prof. Dr.-Ing. Christoph Herrmann
Prof. Sami Kara

Acknowledgments

This book was written during my time as a research associate at the Chair of Sustainable Manufacturing & Life Cycle Engineering at the Institute of Machine Tools and Production Technology, Technische Universität Braunschweig. I owe very special thanks to Prof. Dr.-Ing. Christoph Herrmann, head of the chair and co-director of the institute, who provided the necessary guidance, support, and freedom throughout my stay at the institute. This book would not have been possible without his belief in the success of our collaboration and our valuable discussions, which guided the way and resulted in the key ideas reflected in this work. I would also like to acknowledge the contribution and support from Prof. Sami Kara, Prof. Dr. Kuldip Singh Sangwan, and Prof. Dr.-Ing. Ulrike Krewer during the process of finalizing this book.

Being part of the Sustainable Manufacturing & Life Cycle Engineering research group was a great experience, not just because of the positive and friendly atmosphere and the warm welcome. Representative of the whole group, I would like to thank Dr.-Ing. Sebastian Thiede for the fruitful discussions and review of this book, and Malte Schönemann for the countless hours of exchanging and criticizing ideas and also his support outside the formal process of finishing this manuscript.

Big thanks also go to numerous friends, who accompanied me during the past three years and provided guidance, but also the necessary distraction to keep my head clear. My parents and sister substantially supported me during this time, for which I would like to thank them very much. Undoubtedly, I owe the most to Desirée, for her limitless support.

Braunschweig, Germany
July 2016

Jan Beier

Contents

Abbreviations

AA-CAES	Advanced-adiabatic compressed air energy storage technology
AC	Air-conditioning
AFOLU	Agriculture and Forestry and Other Land Use
ANN	Artificial neural network
CA	Compressed air
CAES	Compressed air energy storage
CHP	Combined heat and power
CO_{2eq}	Carbon dioxide equivalent
c.p.	ceteris paribus
CPP	Critical peak pricing
DCM	Die casting machine
DIHK	Association of German Chambers of Commerce and Industry (German: *Deutscher Industrie- und Handelskammertag*)
DR	Demand response
DSM	Demand side management
ED-CPP	Extreme day critical peak pricing
EDP	Extreme day pricing
EEX	European Energy Exchange
e-KPI	Energy-related key performance indicator
ENPI	Energy performance indicator
ENTSO-E	European Network of Transmission System Operators for Electricity
EV	Electric vehicle
FC	Combining flow element
FD	Dividing flow element
G2V	Grid-to-vehicle
GDP	Gross domestic product
GHG	Greenhouse gas
GUI	Graphical user interface
HVAC	Heating, ventilation, air-conditioning/cooling

i.i.d.	Independent and identically distributed
ILP	Integer linear programming
IPCC	Intergovernmental Panel on Climate Change
ISACOAST-CC	Isobaric adiabatic compressed air energy storage plant with combined cycle
IWF	Institute of Machine Tools and Production Technology
KPI	Key performance indicator
LCA	Life cycle assessment
LM	Load management
MDP	Markov decision process
MES	Manufacturing execution system
MILP	Mixed integer linear programming
MINLP	Mixed integer nonlinear programming
MRR	Material removal rate
MTBF	Mean time between failures
MTTF	Mean time to failure
MTTR	Mean time to repair
NIP	Nonlinear integer programming
OR	Operations research
OTC	Over-the-counter
Phelix	Physical Electricity Index
PLC	Programmable logic controller
PSO	Particle swarm optimization
PV	Photovoltaics
RE	Renewable energy
RTO	Regional transmission organization
RTP	Real-time pricing
SMES	Superconducting Magnetic Energy Storage
TBS	Technical building services
TOU	Time-of-use
TSO	Transmission system operator
UCTE	Union for the Co-ordination of Transmission of Electricity
V2G	Vehicle-to-grid
VRE	Variable renewable energy
WACC	Weighted average cost of capital
WCED	World Commission on Environment and Development
WIP	Work-in-progress
ZONLP	Zero-one nonlinear programing

Symbols

Latin Symbols

a	Production process index (number)
ab	Adjustable binary process index, $ab = 1, \ldots, AB$ (number)
AB	Total number of adjustable binary processes (number)
ac	Adjustable continuous process index, $ac = 1, \ldots, AC$ (number)
AC	Total number of adjustable continuous processes (number)
Av_e	Availability of element e (%)
AvI_t	Average inventory until time t (pieces)
b	Battery index, $b = 1, \ldots, B$ (number)
B	Total number of batteries (number)
$BC_{b,t}$	Battery capacity of battery b at time t (Wh)
bne	Battery index for not fully discharged batteries, $bne = 1, \ldots, BNE$ (number)
BNE	Total number of not fully discharged batteries (number)
bnf	Battery index for not fully charged batteries, $bnf = 1, \ldots, BNF$ (number)
BNF	Total number of not fully charged batteries (number)
$BT_{n,t}$	Block time of process n at time t (min)
c	Production process index (number)
C_n	Capacity of buffer n (products)
CA_t	Compressed air flow at time t (standard cubic meters per minute (Nm^3/min))
$CA_n^{spec.}$	Rate-dependent CA demand of a process (standard cubic meters per product (Nm^3/prod.))
$CASV_{sys.}$	Compressed air system storage volume (cubic meters (m^3))
$Cc_{cs,t}$	Hourly switch count of compressor cs at time t (number)
$CCA_{sys.,t}^{avail.}$	Available compressed air at time t (standard cubic meters (Nm^3))
CEl_t	Cumulative electricity until time t (Wh)
$CP_{n,t}$	Cumulative production of process n until time t (products)

CR_b	Battery max. charge rate (W)
cs	Compressor index, $cs = 1, \ldots, CS$ (number)
CS	Total number of compressors in compressor park (number)
CSS_t	Rolling switch count of compressors at time t (number)
csa	Adjustable compressor index, $csa = 1, \ldots, CSA$ (number)
CSA	Total number of adjustable compressors in compressor park (number)
Ct_n	Cycle time of process n (seconds)
$Cyc_{b,t}$	Number of full battery cycles until time t (number)
$Cyc_b^{x\%}$	Number of full battery cycles until x% initial capacity remains (number)
d	Production process index (number)
$\Delta p_{crit.}$	Pressure threshold to determine compressor park critical (bar)
DCR_b	Battery max. discharge rate (W)
e	System element index, $e = 1, \ldots, E$ (number)
E	Total number of system elements (number)
$EICA_t$	Electric energy intensity of compressed air at time t (watt-hours per standard cubic meter (Wh/Nm^3))
El	Electricity demand/supply (used with sub-/superscript for reference) (W)
$ElAP_{ac,t}$	Adjustment potential of continuous process ac at time t (W)
esf_n	Energy saving factor for process n switch-off (number)
$FL_{n,t}$	Current fill level of buffer n at time t (products)
k	Production process index (number)
l	(Intermediate) product index, $l = 1, \ldots, L$ (number)
L	Total number of (intermediate) products in the system (number)
MED_t	Maximum external demand at time t (W)
$MTBF$	Mean time between failures (s)
$MTTR$	Mean time to repair (s)
MV_t	Monetary value at time t (Euro)
n	Process and buffer number index, $n = 1, \ldots, N$ (number)
N	Total number of processes (number)
nc	All possible on/off combinations index, $nc = 1, \ldots, NC$ (number)
nc_o	Optimal on/off combination (number)
NC	All possible on/off combinations (number)
ncp	Index for a specific on/off control signal combination for binary processes $ncp = 1, \ldots, NCP$ (number)
NCP	All possible on/off combinations for binary processes (number)
$OGDR_t$	On-site generation demand ratio at time t (%)
os	Own electricity supply sources index, $os = 1, \ldots, OS$ (number)
OS	Total number of own electricity supply sources (number)
PHR_n	Process hourly rate (Euro/h)
p_t	CA system pressure at time t (bar)
P_n	Production process n (process)
PrE_l	Embodied energy of product l (Wh)
PrT_l	Entry time of product l into the system (h)

PrV_l	Monetary value of product l (Euro)
$PV_{ac,t}$	Priority value of process ac at time t (1/min)
q	Production process index (number)
r	Production process index (number)
$R_{n,t}$	Production rate of process n at time t, $R_{n,t} = 1/Ct_{n,t}$ (products/s)
$R_n^{max.}$	Maximum production rate of continuous process n (products/s)
s	Electricity supply sources index, $s = 1,\ldots,S$ (number)
S	Total number of electricity supply sources (number)
$SOC_{b,t}$	(Battery) state of charge at time t (Wh)
$SOffT_e$	Switch-off time of element e (s)
$SOnT_e$	Switch-on time of element e (s)
Srt_t	System residence time until time t (min)
SSR_t	Self-sufficiency ratio at time t (%)
$ST_{n,t}$	Starve time of process n at time t (min)
t	Time, $t = 1,\ldots,T$ (min)
T	Total model/simulation time (h)
TEL_t	Target electricity demand (W)
TR_t	Throughput until time t (products)
$TW_{n,t}$	Time window of process n at time t (s)
$TWS_{ac,t}$	Time window share of process ac at time t (s)
ty	Product type index, $ty = 1,\ldots,TY$ (number)
TY	Total number of product types (number)
U_m	Maximum rating of an electric power network (V)
U_n	Nominal rating of an electric power network (V)
V	Volume (cubic meters (m^3))
wt_e	Idle-waiting time of element e (s)

Greek Symbols

α	Scale parameter of Weibull distribution (–)
β	Shape parameter of Weibull distribution (–)
η_b	Battery efficiency of battery b (%)
ι_{CA}	CA target production rate adjustment factor (%)
μ	Mean of normal distribution (%)
ν_e	Binary control signal (produce/idle) for binary process e, $\nu_e = \{0,1\}$ (%)
ξ_b	Self-discharge factor of battery b (%)
σ	Standard deviation of normal distribution (%)
$\phi_{s,t}$	Electricity share of supply source s at time t (%)

Further Frequently Used Subscript and Superscript

The following subscripts and superscripts are used in combination with above symbols. Within this table, the symbol X is used to differentiate between subscript and superscript without further meaning. Note that some elements might be used as sub- and superscript.

$X^{avail.}$	Reference to available, e.g., available compressed air in the system
$X^{avg.}$	Reference to average value (arithmetic mean unless otherwise noted)
$X_{bat.}$	Reference to battery (system) value
$X^{bin.}$	Reference to binary process value
X_{bn}	Reference to bottleneck value
X^{CCA}	Reference to cumulative compressed air
X^{charge}	Reference to charge value
$X^{con.}$	Reference to continuous process value
$X_{conv.}$	Reference to conversion
X_{cop}	Reference to compressor park
$X_{crit.}$	Reference to critical
$X_{cust.}$	Reference to customer
$X^{dem.}$	Reference to demand, e.g., electricity or compressed air
X^{direct}	Reference to direct electricity demand
X_{end}	Reference to end
X^{grid}	Reference to value for grid electricity supply
X^{idle}	Reference to value when process is in state *idle*
X^{in}	Reference to inflow/input
$X^{init.}$	Reference to initial amount, e.g., initial available compressed air in the system
X^{loss}	Reference to lost amount, e.g., lost compressed air
X^{max}	Reference to maximum value
X^{min}	Reference to minimum value
X^{off}	Reference to value when process is in state *off*
X^{out}	Reference to outflow/output
X^{own}	Reference to value for own electricity supply
$X^{prod.}$	Reference to value when process is in state *produce*
$X^{sched.}$	Reference to scheduled value
$X^{selfdis.}$	Reference to self-discharge value
X^{solar}	Reference to value for solar electricity supply
$X^{spec.}$	Reference to rate-dependent value
X_{start}	Reference to start
X^{supply}	Reference to supply
$X_{sys.}$	Reference to value for (total) system
X^{target}	Reference to target value
X^{total}	Reference to value for total electricity supply

$X^{turn\ off}$ Reference to value when process is in state *switching off*
$X^{turn\ on}$ Reference to value when process is in state *switching on*
$X^{var.}$ Reference to variable part of value
X^{wind} Reference to value for wind electricity supply

Metric Prefixes

Note that units might be combined with the following metric prefixes:

Prefix	**Name**	**Factor**
T	tera	1 000 000 000 000
G	giga	1 000 000 000
M	mega	1 000 000
k	kilo	1 000
c	centi	0.01
m	milli	0.001

List of Figures

List of Tables

Chapter 1
Introduction

More than a quarter century ago, the World Commission on Environment and Development (WCED) coined the concept of *sustainable development* with their report "Our common future" (World Commission on Environment and Development 1987). At least since then, sustainable development has been of increasing interest within research, politics, industry, and society in general. A central goal stated by the Intergovernmental Panel on Climate Change (IPCC) for achieving sustainable development is the mitigation of climate change and thus the reduction of greenhouse gas (GHG) emissions, mainly carbon dioxide (IPCC 2014c). Although various national and international policies and goals have been defined to limit and/or reduce greenhouse gases (e.g., the Kyoto Protocol), emissions increased twice as fast between 2000 and 2010 compared to any other 10-year period since the last 45 years. In 2010, global GHG emissions were estimated to be 49.5 billion tons carbon dioxide equivalent (Gt CO_2eq), higher than ever before. From these 49.5 Gt CO_2eq, the single largest contributing sector as defined by the IPCC was *Electricity and Heat Production* with 25 %, followed by *Agriculture and Forestry and Other Land Use* (23 %) and *Industry* with 18 %. As electricity and heat production are not independent from their respective end use, allocating these emissions to causing sectors pushes industry to the top with 28.6 % of global emissions. Figure 1.1 shows the distribution of emissions across all sectors, split by direct and indirect (allocated electricity and heat production) emissions.

1.1 Renewable Energy Sources for Sustainable Development

Sustainable development usually encompasses a variety of objectives such as combining economic growth with social and economic goals (IPCC 2014c). As such,

J. Beier, *Simulation Approach Towards Energy Flexible Manufacturing Systems*, Sustainable Production, Life Cycle Engineering and Management, DOI 10.1007/978-3-319-46639-2_1

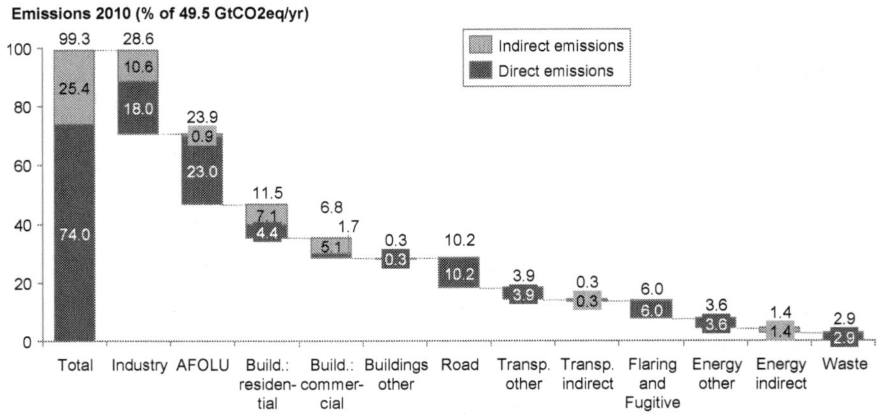

Fig. 1.1 Distribution of 2010 GHG emissions across sectors, split by direct and indirect emissions (AFOLU: Agriculture and Forestry and Other Land Use, note that values might not add up to 100 % due to rounding, data from IPCC (2014c), own illustration)

strategies and policies to foster sustainable development cannot be pursued separately. In the context of climate change mitigation and emission reduction, decoupling economic growth from resource and energy demand is a central concern (UNEP 2011; IPCC 2014a). From 1960 to 2008, GHG emissions grew, on average, 0.73 % for every one percent increase in gross domestic product (GDP) per capita, indicating that economic growth increases faster than emissions, but also that full decoupling remains challenging (IPCC 2014a). Further, adjusting for consumption-based emissions (accounting for trade) compared to territorial emissions used above, GHG emission growth per GDP growth (elasticity) is even higher for industrialized countries (IPCC 2014a).

As noted before, GHG emissions are closely linked to energy demand. Figure 1.2 shows the energy intensity per GDP and per capita for Germany from 1990 to 2013. Albeit a substantial decrease in emissions per Euro GDP, emissions per capita decreased much slower due to increasing economic welfare. Further, considering electricity demand, electrical energy demand per GDP decreased over the same period, while electricity per capita increased above 1990 levels. In the same time, specific CO_2 emissions per kWh electricity decreased, resulting in a decrease in emissions per capita from electricity despite higher total electricity demand. Nonetheless, per kWh electricity and per capita CO_2 emissions have been increasing since 2010, highlighting the need for (further) electricity related emission reduction.

In its 2014 report, the IPCC highlighted several options to reduce GHG emissions from energy supply (IPCC 2014b). One option is the deployment of low emission energy and electricity generation technologies such as renewable energy (RE). Among other countries, Germany has enacted a set of policies, summarized under the term *Energiewende* (lit. "energy turnaround"), to restructure the national energy

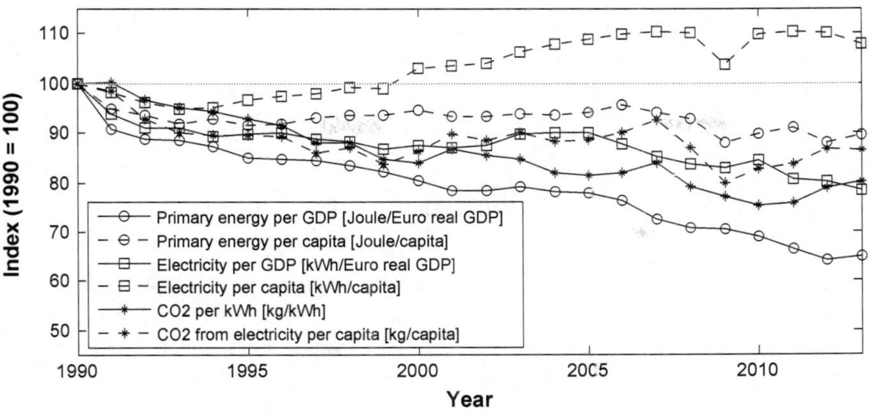

Fig. 1.2 Energy intensity and electricity per Euro real gross domestic product (GDP) and per capita, specific CO_2 emissions per kWh electricity generation and per capita emissions in Germany from 1990 to 2013 (data from AG Energiebilanzen e.V. (2015b), Icha (2014), own calculations and illustration)

supply and to reduce carbon emissions. Until 2050, the country aims at reducing GHG emissions by 80 to 95 % compared to 1990 and wants to establish a nuclear-free energy generation. Complementary to these goals is a targeted 20 % reduction of primary energy requirements until 2020 and 50 % until 2050 compared to 2008. Further, renewable energy share should increase to 60 % of gross final energy demand and 80 % of gross electricity demand in 2050 (Bundesministerium für Wirtschaft und Energie (BMWi) 2014). Figure 1.3 shows Germany's electricity mix from 1990 to 2014 split by renewable and remaining (mainly nuclear and fossil fuel based) electricity generation. From 627.8 TWh gross electricity supply in 2014, 162.5 TWh or 25.9 % was supplied through renewable generation, which has been constantly increasing within the last years (AG Energiebilanzen e.V. 2015a). In sum-

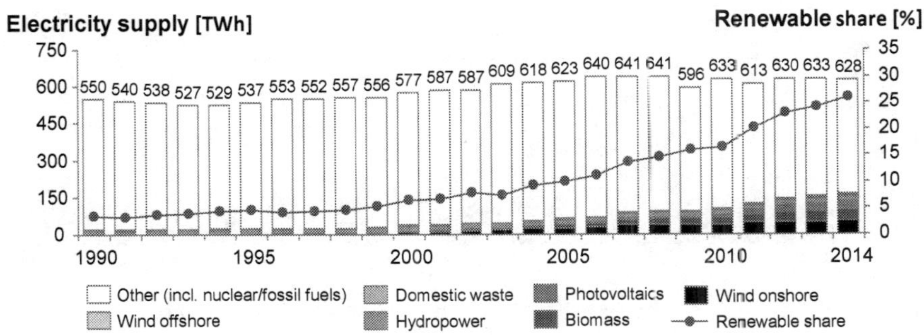

Fig. 1.3 German renewable electricity generation, remaining gross electricity demand, and corresponding renewable share (*right axis*) from 1990 to 2014 (data from AG Energiebilanzen e.V. (2015a), own calculations and illustration)

mary, renewable electricity generation capacity deployment and stated goals within the *Energiewende* are in line with recommendations from the IPCC and contribute towards GHG emission reduction and sustainable development.

1.2 Towards Energy Flexible Manufacturing Systems

Large-scale deployment of RE sources for electricity generation provides a special challenge when variable renewable energy (VRE) sources start to supply a substantial share of electricity supply (Deutsche Energie-Agentur GmbH (dena) 2010; International Energy Agency 2007; National Renewable Energy Laboratory 2013). VRE is characterized by intermittent, non-controllable energy availability, i.e., generation is non-dispatchable (as opposed to, e.g., fossil fuel plants from which electricity output can be controlled). Especially wind and solar generation are influenced by weather conditions, which can, for example, include seasonal changes in solar radiation or clouds which reduce solar intensity locally within seconds to minutes. In addition, generation forecasts underlie inherent forecasting errors. Consequently, power systems (supply, distribution, demand) need to find measures to integrate volatile, uncertain, and non-dispatchable electricity supply from VRE.

In addition to VRE supply, the demand side for electricity (e.g., households, industry) exhibits volatility itself, combined with short-term inelastic demand (Aggarwal et al. 2009; Moghaddam et al. 2011; Backlund et al. 2012; Torriti et al. 2010). Figure 1.4 illustrates the hourly wind and solar generation, residual load and wind plus solar share of total load in Germany during June 2013. Total system load (the sum of all curves) exhibits a recurring pattern. Demand is highest from Monday (e.g., June 3) to Friday, reduced on Saturday and lowest on Sunday. Further, demand is high during day and reduced to a base-load demand for a few hours at night. Looking into intraday variations, residential customers cause, especially on weekends, morning/noon and evening peaks, which is characteristic for European countries (Li and Flynn 2006). On the other side, wind and solar generation are independent

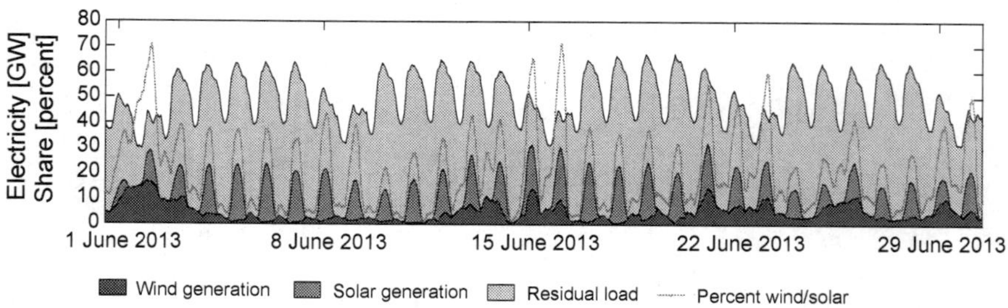

Fig. 1.4 Wind and solar generation and residual load Germany, wind plus solar share from total load, during June 1 to 30, 2013, hourly resolution (data from ENTSO-E (2014), EEX (2014a), EEX (2014b), own calculations and illustration)

of demand (and vice versa): solar generation follows a steady daily increase and decrease between sunrise and sunset with varying intensity between days due to, e.g., clouds. Wind fluctuates with substantial hour-to-daily variations, e.g., up to 15 GW output on June 2, 15, and 21, and long periods of low output in between. As a result of inelastic demand and VRE supply in combination with missing large-scale energy/electricity storage, residual generation, e.g., via dispatchable fossil fuel power plants, needs to ensure that supply matches demand. As VRE supply and demand are not directly related, remaining supply has to be increasingly flexible (compare to wind plus solar share of total demand in Fig. 1.4). One effect is more volatile and even negative prices. For example, Phelix (Physical Electricity Index) spot prices were negative on June 16 from 12 am to 5 pm, with a price of −100 Euro per MWh between 2 pm and 4 pm (i.e., buying/selling one MWh of electricity would generate an income/cost of 100 Euro) (EPEX SPOT 2014). Solar and wind generation share of total load was 71 % (2–3 pm) and 67 % (3–4 pm), 71 % being the highest value in June 2013 (ENTSO-E 2014; EEX 2014a, b).

Another difference of RE sources compared to conventional (fossil fuel) based generation is their geographically diversified and fixed nature. For example, a fossil fuel plant, such as coal or gas fired plants, can be built (within constraints) where required and planned, while, for some RE sources such as wind and solar, generation needs to take place where RE resources are available. This might result in increased electricity transportation requirements, e.g., for the case of Germany, northern off- and onshore wind resources are distant from large industry hubs in western and southern Germany. Another result is geographically decentralized generation, i.e., wind turbines are spread throughout the country, and a significant amount of solar panels is installed on domestic, commercial and industrial rooftops. Decentralized RE generation has therefore been increasing in recent years (IPCC 2014b). A summary of main, previously discussed differences between conventional and renewable energy generation can be found in Table 1.1.

Table 1.1 Comparison between conventional and renewable energy generation

	Conventional energy generation	Renewable energy generation
Variability of output	Stable output level	Second, minute, hourly, daily and seasonal changes
(Geographic) location	Built where desired (within constraints)	Generation needs to take place where available
Centrality	Commonly central/large-scale	Commonly decentralized/small- to medium-scale
Dispatchability	Ramp-up and -down feasible	Only curtailment (reduction) possible
Plannability	Plannable, aside from breakdowns	Planning relies on (stochastic) forecasting plus impact from breakdowns

Summarized, the overall challenge is to match electricity supply and demand considering geographic and temporal mismatches, for which several options exist. Among them are advanced generation forecasting, improved generation reserves management, flexible markets and generation technologies, energy storage and demand response (DR) (National Renewable Energy Laboratory 2013). Energy storage is a promising approach for VRE integration, but remaining capacity for established technology (e.g., pumped hydro) is limited, at least in Germany (Germany Trade and Invest 2015). New technologies, such as hydrogen or battery storage, still face economic challenges to provide large-scale energy storage (Pieper and Rubel 2011). While more flexible generation focuses on the supply side and energy storage aims at overcoming intertemporal differences, DR focuses on a more flexible electricity demand. Implementing DR measures in the context of increasing VRE penetration can help to increase grid stability and can successfully reduce required reserve capacities or energy storage (Deutsche Energie-Agentur GmbH (dena) 2010; National Renewable Energy Laboratory 2013).

In order to flexibilize demand, the main electricity-demanding entities need to be identified. As such, industry was the largest electricity-demanding sector with 43.5 % of total electricity demand in Germany in 2012 (AG Energiebilanzen e.V. 2013). Further, considering its structure and other parameters (e.g., technology), industry offers large potential for DR methods (Deutsche Energie-Agentur GmbH (dena) 2010; Paulus and Borggrefe 2011). Within this context, energy flexibility might be used to describe the potential of production/manufacturing systems and industry in general for implementing DR methods. Therefore, energy flexibility of industry can contribute to an efficient integration of VRE into a power system.

1.3 Objectives and Structural Outline

The described challenges of renewable electricity supply deployment and integration to foster development towards a sustainable society constitute the background and motivation for this book. Figure 1.5 summarizes five (macro-)factors (1a–5a) that have been described in the previous sections which need consideration to successfully achieve a transition towards increased VRE electricity supply, especially in Germany. Considering these five factors, a top-down research question can be formulated:

How can electricity demand of industry be aligned to (decentralized) variable renewable electricity supply?

However, no general study on integrating VRE is targeted (among other reasons, extensive and detailed studies are already available, see, e.g., Deutsche Energie-Agentur GmbH (dena) (2010)), but development of more specific (technical) measures and tools. Therefore, a second, bottom-up research question is formulated, which is based on additional five factors (Fig. 1.5, 1b–5b). In particular, a changing business/operating environment for industry, i.e., volatile energy prices which are

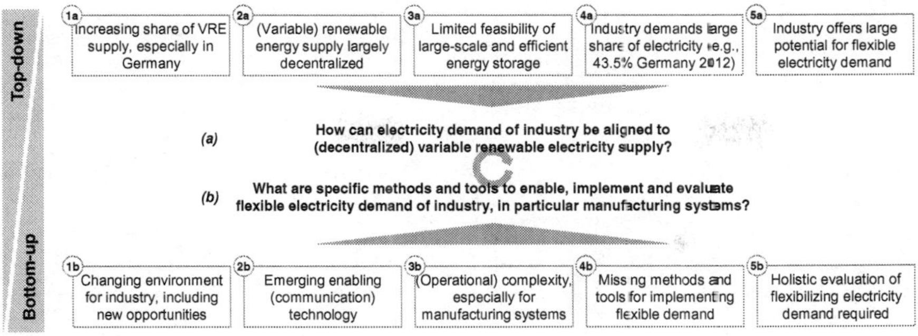

Fig. 1.5 Derivation of *top-down* and *bottom-up* research question (own illustration)

passed through to customers, provides new opportunities (and challenges) for companies to participate and profit from a structurally changing electricity market (1b) (Lorenz et al. 2012). Within this context, required technology that enables communication between suppliers, customers, and grid operators becomes widely available (2b in Fig. 1.5, sometimes included in terms such as smart grid and industry 4.0, which has led to the term electricity 4.0) (Maubach 2015). Both factors combined offer new opportunities for industry, which have not been available before and have therefore not (yet) been pursuit. However, operating complexity, especially for (interlinked) manufacturing systems, provides additional challenges for flexible electricity demand of manufacturing companies (3b in Fig. 1.5, will be detailed in the following chapters). As such, specific methods and tools which lead to successful implementation are required, but are not (yet) available, mainly due to the fact that flexible electricity demand was not of special interest for industry (4b, compare also to first two factors 1/2b, will be substantiated in the following). Related to specific tools for implementation are methods to holistically evaluate actions that target a more flexible electricity demand of industry (5b). This evaluation includes operational, economic and ecological dimensions to establish a sound basis for decision-making. In particular, impacts caused by flexible electricity demand need to be quantified to avoid suboptimization without considering dynamic/second-order effects on operational, economic, and ecological performance. Considering these additional five factors, the following bottom-up research question can be formulated:

What are specific methods and tools to enable, implement and evaluate flexible electricity demand of industry, in particular manufacturing systems?

Both research questions (top-down and bottom-up) are closely related to each other. However, answering the bottom-up research questions will automatically contribute towards answering the top-down research question. Therefore, as a contribution to ease integration of VRE into an existing electricity supply and demand system, an approach is presented, with a focus on specific methods and tools, to (a) design, improve, implement, and evaluate energy flexibility of manufacturing sys-

	Chapter	Content	Abstraction level	Objective
1	**Introduction**	▪ Sustainable development and energy demand ▪ Renewable energy integration ▪ Objective and outline	Global perspective	To describe the overall context, objective and structure.
2	**Manufacturing systems and variable renewable electricity supply**	▪ Manufacturing systems ▪ Electricity supply ▪ Integrating renew. energy	Established theory, state of the art	To provide the required theoretical background.
3	**Existing approaches in the field of energy flexible manufacturing systems**	▪ Localization of approaches ▪ Classification of approaches ▪ Evaluation of approaches ▪ Derivation of research demand	Existing, relevant approaches	To review existing research and derive research demand.
4	**Manufacturing system real-time energy flexibility control and improvement**	▪ Concept objective/structure ▪ Concept detailed description ▪ Application cycle	Description of own concept, independent from implementation	To present concept for energy flexibility improvement and control
5	**Prototypical implementation**	▪ Implementation options ▪ Required background ▪ Software simulation model	Concept implementation into general software tool	To demonstrate feasibility of concept and to supply general toolbox.
6	**Example application**	▪ Example use-cases ▪ Background and target ▪ Modeling and implementation ▪ Evaluation of results	Existing use-cases implemented in tool	To show applicability of concept with real-life data input.
7	**Summary, critical review and outlook**	▪ Results and accomplishments ▪ Discussion of assumptions and influencing factors ▪ Future research direction	Aggregation of findings and relation to greater context	To summarize findings, criticize and provide future leads.

Fig. 1.6 Structural outline (own illustration)

tems and (b) control manufacturing systems under energy flexibility goals. Figure 1.6 summarizes the structural outline including a brief description of main content and objectives.

References

AG Energiebilanzen e.V. (2013). Anwendungsbilanzen für die Endenergiesektoren in Deutschland in den Jahren 2011 und 2012 mit Zeitreihen von 2008 bis 2012. Technical report, Berlin.

AG Energiebilanzen e.V. (2015a). Bruttostromerzeugung in Deutschland ab 1990 nach Energieträgern. Technical report, Berlin. http://www.ag-energiebilanzen.de/index.php?article_id=29&fileName=20151112_brd_stromerzeugung1990-2014.pdf.

AG Energiebilanzen e.V. (2015b). Effizienzindikatoren. Retrieved August 17, 2015, from http://www.ag-energiebilanzen.de/38-0-Effizienzindikatoren.html.

Aggarwal, S. K., Saini, L. M., & Kumar, A. (2009). Electricity price forecasting in deregulated markets: A review and evaluation. *International Journal of Electrical Power & Energy Systems*, *31*(1), 13–22.

Backlund, S., Thollander, P., Palm, J., & Ottosson, M. (2012). Extending the energy efficiency gap. *Energy Policy*, *51*, 392–396. http://linkinghub.elsevier.com/retrieve/pii/S0301421512007240.

Bundesministerium für Wirtschaft und Energie (BMWi) (2014). Zweiter Monitoring-Bericht "Energie der Zukunft". Technical report, Berlin.

Deutsche Energie-Agentur GmbH (dena) (2010). dena Grid Study II. Integration of Renewable Energy Sources in the German Power Supply System from 2015–2020 with an Outlook to 2025. Technical report, Berlin.

EEX (2014a). Transparency in Energy Markets - Actual Solar Power Generation (table). Retrieved March 29, 2014, from http://www.eex-transparency.com/homepage/power/germany/production/usage/actual-solar-power-generation-/actual-solar-power-generation-table-.

EEX (2014b). Transparency in Energy Markets - Actual Wind Power Generation (table). Retrieved March 29, 2014, from http://www.eex-transparency.com/homepage/power/germany/production/usage/actual-wind-power-generation/actual-wind-power-generation-table-.

ENTSO-E (2014). Hourly load values for a specific country for a specific month (in MW). Retrieved March 31, 2014, fromhttps://www.entsoe.eu/db-query/consumption/mhlv-a-specific-country-for-a-specific-month.

EPEX SPOT (2014). EPEX SPOT SE: Day-Ahead Auction. Retrieved March 29, 2014, from http://www.epexspot.com/en/market-data.

Germany Trade and Invest (2015). The energy storage market in Germany. Technical report, Köln.

Icha, P. (2014). Entwicklung der spezifischen Kohlendioxid-Emissionen des deutschen Strommix in den Jahren 1990 bis 2013. Technical report, Umweltbundesamt, Dessau-Roßlau.

International Energy Agency (2007). Contribution of Renewables to Energy Security. Technical report, Paris.

IPCC (2014a). Climate change 2014: Mitigation of climate change. Working group III contribution to the IPCC 5th assessment report: Chapter 5: Drivers, trends and mitigation. In *Climate Change 2014: Mitigation of Climate Change Working Group III Contribution to the IPCC 5th Assessment Report*. Cambridge, New York.

IPCC (2014b). Climate change 2014: Mitigation of climate change. Working group III contribution to the IPCC 5th assessment report: Chapter 7: Energy systems. In *Climate Change 2014: Mitigation of Climate Change Working Group III Contribution to the IPCC 5th Assessment Report*. Cambridge, New York: Cambridge University Press.

IPCC (2014c). Climate change 2014: Mitigation of climate change. Working group III contribution to the IPCC 5th assessment report: Introductory chapter. In *Climate Change 2014: Mitigation of Climate Change Working Group III Contribution to the IPCC 5th Assessment Report*. Cambridge, New York.

Li, Y., & Flynn, P. (2006). Electricity deregulation, spot price patterns and demand-side management. *Energy*, *31*(6–7), 908–922. http://linkinghub.elsevier.com/retrieve/pii/S0360544205000502.

Lorenz, S., Putz, M., & Schlegel, A. (2012). Energieeffizienz 2.0: Neue Geschäftsmodelle auch für die Industrie. *ZWF Zeitschrift für wirtschaftlichen fabrikbetr*, *107*(9), 599–602.

Maubach, K.-D. (2015). Strom 4.0. http://link.springer.com/10.1007/978-3-658-08613-8.

Moghaddam, M. P., Abdollahi, A., & Rashidinejad, M. (2011). Flexible demand response programs modeling in competitiveelectricity markets. *Applied Energy*, *88*(9), 3257–3269. http://dx.doi.org/10.1016/j.apenergy.2011.02.039.

National Renewable Energy Laboratory (2013). Integrating Variable Renewable Energy: Challenges and Solutions. Technical report, National Renewable Energy Laboratory, Golden, CO.

Paulus, M., & Borggrefe, F. (2011). The potential of demand-side management in energy-intensive industries for electricity markets in Germany. *Applied Energy*, *88*(2), 432–441.

Pieper, C., & Rubel, H. (2011). Revisiting energy storage: There Is a Business Case. Technical report, The Boston Consulting Group.

Torriti, J., Hassan, M. G., & Leach, M. (2010). Demand response experience in Europe: Policies, programmes and implementation. *Energy*, *35*(4), 1575–1583. http://dx.doi.org/10.1016/j.energy.2009.05.021.

UNEP. (2011). *Decoupling Natural Resource Use and Environmental Impacts from Economic Growth*. United Nations Environment Programme, Paris.

World Commission on Environment and Development (1987). Our common future. Technical report, World Commission on Environment and Development, Oxford. http://dx.doi.org/10.1002/jid.3380010208.

Chapter 2
Manufacturing Systems and Variable Renewable Electricity Supply

After formulating the central research question(s), this chapter provides necessary background and theory required for the development of the manufacturing system energy flexibility control and improvement concept. As a starting point, Sect. 2.1 presents the background on manufacturing systems, their management and control as well as energy demand of processes and systems, with a special focus on energy flexibility. The next section starts with definitions related to energy supply and provides additional background on conventional electricity generation in the context of mitigating climate change. In the following, background on renewable electricity generation, especially decentralized generation, is given. As a combining element of demand and supply, electricity distribution, markets and related challenges regarding grid stability are discussed. After highlighting electricity demand and supply, options to integrate RE into an existing electricity system are discussed. Integration options are categorized into energy storage and demand side management (DSM) strategies to flexibilize demand. The chapter closes with a brief summary and conclusion.

2.1 Manufacturing Systems

This section starts with necessary background on manufacturing systems, including definitions, management and control strategies, energy demand and energy flexibility of manufacturing systems.

2.1.1 Definition and Classification

(Industrial) production can be defined as the creation of outputs (products) from material and non-material inputs (production factors) according to specific technical

J. Beier, *Simulation Approach Towards Energy Flexible Manufacturing Systems*, Sustainable Production, Life Cycle Engineering and Management, DOI 10.1007/978-3-319-46639-2_2

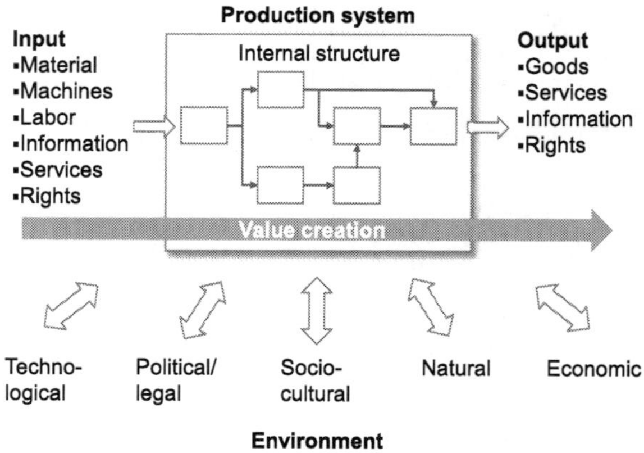

Fig. 2.1 Production system and its environment (based on Günther and Tempelmeier 2005, p. 2; Dyckhoff 2006, p. 5; Dyckhoff and Spengler 2010, p. 4, translated from German, own illustration; compare also to Wiendahl 2014; Westkämper 2006; Zäpfel 2001). Reprinted with permission from Günther and Tempelmeier (2005), Dyckhoff (2006) and Dyckhoff and Spengler (2010)

procedures (Günther and Tempelmeier 2005, p. 6, translated from German, compare also to Zäpfel 2001, p. 1).[1] In other words, production can be regarded as a transformation process to create value (Dyckhoff and Spengler 2010). *Manufacturing*, on the contrary, can be defined as "the transformation of materials and information into goods for the satisfaction of human needs" (Chryssolouris 1992, p. 2). Similar to production, a central aspect is the transformation of inputs to outputs through a transformation process to create value. Further, and more specific, a *manufacturing process* might be defined as "the use of one ore more physical mechanisms to transform a material's shape and/or form" (Chryssolouris 1992, p. 34). In general, an industrial manufacturing process transforms material objects to products using labor and machines (Günther and Tempelmeier 2005; compare also to Westkämper 2006 for definitions/delimitation of and background on production and manufacturing). Within the following, the terms manufacturing process and production process are both referring to above definition with relation to industrial production (as opposed to, e.g., production of intangible products, such as producing a movie). Consequently, the terms *production* and *manufacturing* are used synonymously in the following unless otherwise noted (e.g., in a context where an original reference uses the term production).

A *production system* encompasses several (similar or different) production processes and related infrastructure to enable material and information flow between processes (Dyckhoff and Spengler 2010; Westkämper 2006). Figure 2.1 shows the interaction of a production system with its environment: a production system with several subsystems/processes creates value by transforming inputs to outputs, and is embedded into and interacts with a broader technological, political/legal, sociocultural, natural and economic environment (Günther and Tempelmeier 2005; Dyckhoff and Spengler 2010; Dyckhoff 2006; Wiendahl 2014; Westkämper 2006).

[1] The term *production* is sometimes also used in the context of electricity production. Within this book, the term *production* refers to above definition, i.e., the creation of products, while *generation* or *electricity generation* is used in the context of energy conversion (unless otherwise noted).

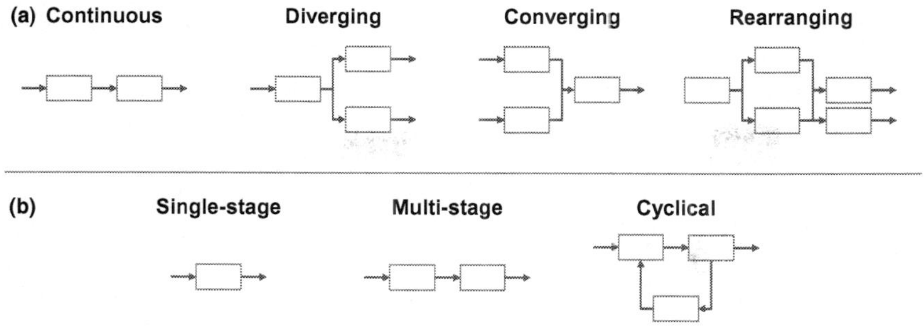

Fig. 2.2 Generic production system material flow structures (**a**) and system stages/network structure (**b**) (based on Dyckhoff and Spengler 2010, p. 22, translated from German, own illustration with additions; see also Schuh and Stich 2012; Dyckhoff 2006). Reprinted with permission from Dyckhoff and Spengler (2010)

Production system typologies can be classified into input, output, and throughput-oriented typologies (see, e.g., Dyckhoff and Spengler 2010). The throughput-oriented type is mainly used to classify and describe production systems within the following as this type focuses on the internal structure of a production system, which is of central relevance for developing an energy flexibility concept. According to the throughput-oriented typology, a production system is characterized according to natural or engineering processes, material flow structure (Fig. 2.2a), system stages/network structure (Fig. 2.2b), repetition type, and spatial alignment of production units (Dyckhoff and Spengler 2010). Different organizational production structures (e.g., job shop production, production line) are, for example, described in Wiendahl (2014).

Another perspective is offered by considering the factory as an entity with interdependent elements connected by material, energy, and information flows. Figure 2.3 shows a holistic understanding of a factory (Herrmann et al. 2014; Thiede 2012; Posselt 2016). The manufacturing system is the central (value-adding) part of the factory. However, in order to fulfill required tasks, technical building services (TBS) are

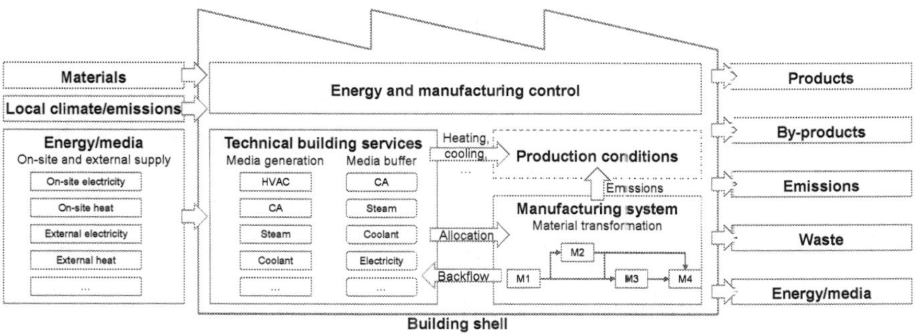

Fig. 2.3 Holistic factory understanding (based on Herrmann et al. 2014; Thiede 2012; Posselt 2016, own illustration)

required to supply media to manufacturing processes and machines, and to process media that flows back from manufacturing activities. TBS include, for example, compressed air (CA) and steam generation. Some media types can be stored before they are required in manufacturing processes. A special role is fulfilled by heating, ventilation and air conditioning/cooling (HVAC), which ensures that desired production conditions (e.g., temperature, air quality) are maintained. These conditions are influenced, aside from outside climate and emission levels, by manufacturing activities. Further, an energy and manufacturing control is used to influence activities (manufacturing, media generation) towards a desired target state. The factory is surrounded by a building shell, which determines the impact of local climate and emissions and influences the emissions released by the factory into the environment. Further, material and energy/media flows are entering the factory, while (by-)products and energy/media flows exit the factory, as well as unwanted flows such as emissions and waste.

2.1.2 Manufacturing System Management and Control

A manufacturing system can be understood as a control cycle, in which the *production management* fulfills the controller function (Dyckhoff and Spengler 2010, Fig. 2.4). While this approach considers a (general) management system separated into strategic, tactical, and operational management (also related to long-, mid-, and short-term planning), (direct) production control and its specific tasks (e.g., resource

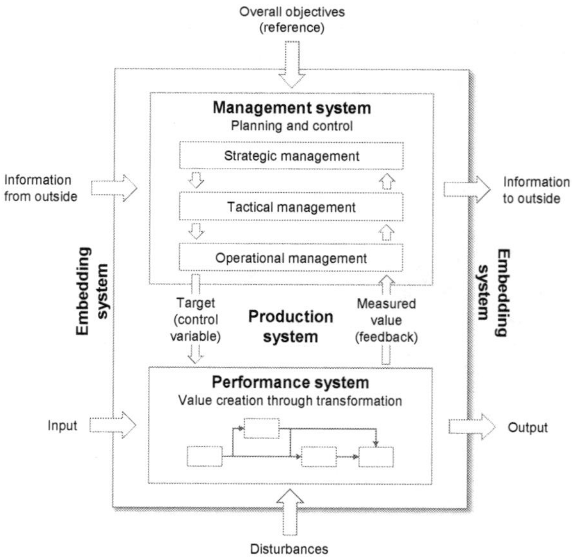

Fig. 2.4 Production management control circle (based on Dyckhoff 2006 and Dyckhoff and Spengler 2010, translated from German, own illustration, similar approach in Wiendahl 2014)

allocation) are also summarized under the term production planning and control (Schuh 2006). The production management aims at planning and controlling the physical performance system according to overall objectives, which are the reference input. Disturbances from embedding systems (the environment, compare also to Fig. 2.1) as well as internal disturbances (i.e., deviation from production schedules due to process failures) require a frequent adjustment of control variables and receiving of feedback to evaluate the effectiveness of control actions.

Depending on its planning horizon, production management can be classified into *strategic* (approx. five years), *tactical* (approx. one to five years), and *operational* (up to one year) production management (Dyckhoff and Spengler 2010; other references consider slightly different planning horizons, e.g., the strategic planning horizon in Wiendahl 2014 is ten to fifteen years or even longer). Typical tasks include

- Strategic production management: production location planning, target and strategy definition, maintaining competitiveness, research and development (products and production processes/technology).
- Tactical production management: layout planning, mid-term production program and capacity planning, technology management.
- Operational production management: short-term production scheduling, material disposition, and production sequence planning.

The operational production management constitutes the main (control) interface to the *performance system* (production system) (Zäpfel 2001). Further, the performance system itself has also several (internal) control cycles to accomplish the required physical transformation (production planning and control).

In order to understand and develop options for influencing manufacturing systems towards improved integration of VRE, the manufacturing process and system energy/electricity demand side is described in the next section.

2.1.3 Energy Demand of Manufacturing Processes and Systems

Manufacturing processes require energy to perform a transformation from inputs into outputs. Typical energy carriers demanded by manufacturing processes are, for example, electricity, compressed air, cooling, and heating[2] (Thiede 2012; Herrmann et al. 2014; Posselt 2016). For the case of electricity demand, several models have been proposed for decomposing and ultimately predicting electricity demand of processes (see, e.g., Gutowski et al. 2006; Kara and Li 2011; Apostolos et al. 2013).

A generic example, e.g., observable for material removal processes such as grinding, turning, or milling, can be found in Fig. 2.5. Electrical power demand over time (left graph) is characterized by different components. These components are demanding electricity depending on current state of the machine, which, in turn,

[2]Note that mentioned energy carriers can be transformed into each other, e.g., compressed air might be generated using electricity as input.

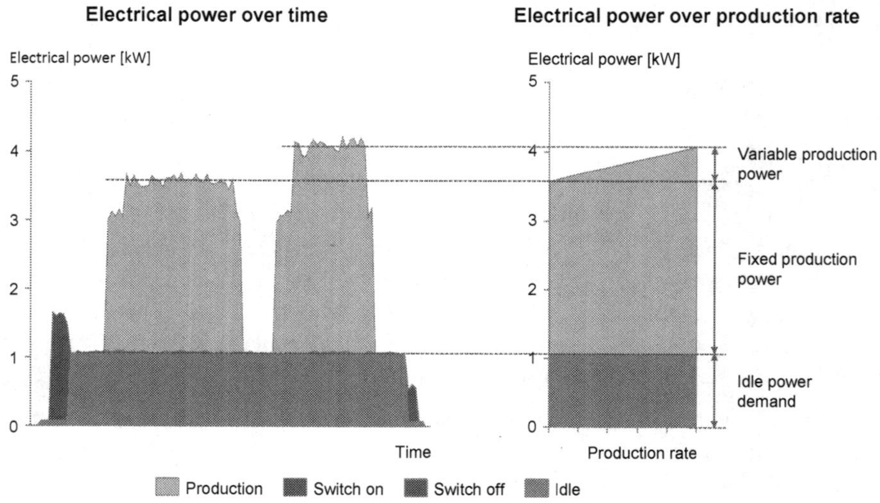

Fig. 2.5 Example manufacturing process energy demand over time and rate (based on Gutowski et al. 2005, 2006; Hesselbach et al. 2008; Kara and Li 2011; Herrmann and Thiede 2009; Weinert 2010; Thiede 2012; Apostolos et al. 2013; Posselt 2016; Duflou et al. 2012, own illustration)

depends on the current task(s) which needs to be fulfilled to complete a desired production sequence. A constant base-load demand, e.g., originating from turned-on controls, is observable throughout the process. Further, switching the machine on and off requires a certain amount of energy (and time). After the machine has been switched on, an idle electricity demand is required by supporting processes such as coolant pumps or a lubrication system (idle demand can include previously mentioned base-load demand, which is assumed in the example in Fig. 2.5). Two production sequences are shown: both sequences first demand additional power to, e.g., accelerate a spindle or tool, followed by a further increase in power demand when actual production (e.g., material removal) takes place. The second production cycle exhibits higher energy demand and a shorter duration. Note that the graph shows a generic example, demand peaks during, e.g., electrical motor switch-on and additional stochastic components are not illustrated (for more detail on classifying and characterizing energy demand of stochastic components such as chip conveyors see, e.g., Popp and Zäh 2014).

The right graph in Fig. 2.5 illustrates electricity demand for different performance levels, i.e., production rate (similar to, e.g., Gutowski et al. 2005). Total electricity demand is split into base-load power/idle demand and production power. Production power can be changed to a certain extent, i.e., by increasing production rate such as material removed per time unit. Several other manufacturing machine energy demand profiles exist, which are influenced by the machine (type), technology and process (for a method to decompose energy demand using different *Energy Blocks* see Weinert 2010 and Weinert et al. 2011). However, state-based energy demand modeling, using multiple states, also in combination with continuous factors, is widely used for energy

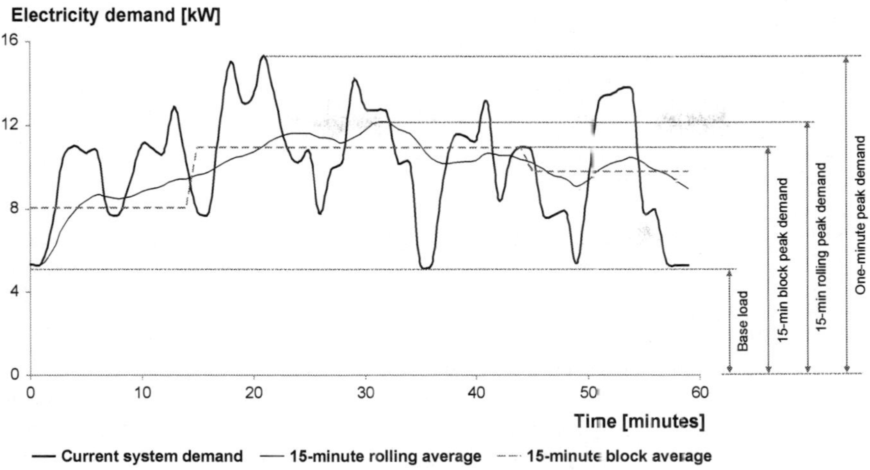

Fig. 2.6 Example manufacturing system energy demand profile (own illustration)

demand modeling of single processes. Further, state-based energy demand can also be extended to other energy carriers than electricity, for example, compressed air (Thiede 2012).

On a manufacturing system level, energy and electricity demand in particular is the sum of electricity demand of individual components. Production sequence and material flow in general as well as (random) disturbances result in highly dynamic demand profiles. Figure 2.6 illustrates an aggregated demand profile of a manufacturing system. The profile has been derived by adding electricity demand from five machines. These five machines are assumed to exhibit electricity demand profiles similar to Fig. 2.5 (only the two different illustrated production process profiles are included, with different idle times between production cycles). Overall, a highly dynamic demand profile is obtained, with four distinctive power levels:

1. System demand is characterized by a base-load (minimum load throughout time period).
2. A 15-minute maximum block average can be calculated, which might be cost relevant (*demand rate*, relevant at the metering point for, e.g., a factory, and dependent upon contract terms).
3. Alternatively, a 15-minute maximum rolling average can be calculated (Measurlogic 2015). Although calculating block averages is more common, utilizing a rolling average and determining a maximum throughout a given period is a more conservative indicator, as a rolling average always includes all block average values and thus a maximum rolling average value is always at least as high as the maximum block average value.
4. The actual (technical) peak demand can be significantly higher (especially relevant in a VRE context where direct demand/supply matching is pursued, and, e.g., for autarkical systems where peaks have to be directly compensated). The

illustrated case uses a one-minute sample rate, which omits technical peaks that can only last for sub-second periods. However, on a system level, these peaks might not be measurable as they are usually small compared to total system load (e.g., a peak from starting an electric drive compared to total load from a manufacturing line or whole factory) and as a total system might absorb small peaks, e.g., due to existing capacitance.

Stochastic influences observable on a single-process level are less significant on an aggregated level. On an overall factory level, several additional energy demanding components can be identified, most notably TBS and HVAC. Especially HVAC energy demand can exhibit strong dynamics and stochastic behavior, as energy demand is dependent on, e.g., process emissions and (stochastic) climate influences. Due to the focus on electricity in the context of VRE integration, discussion of additional flows and energy carriers is not detailed here. Detailed background can be found in references from this subsection and in particular in Hesselbach et al. (2008), Thiede (2012), Posselt (2016) and Herrmann et al. (2014). For the case of factory-wide electricity demand, distinctive demand patterns can usually be observed, similar to grid demand patterns (cf. Fig. 1.4). Factory demand patterns are mainly driven by production activities and their sustaining processes (TBS, HVAC, administration). For example, weekday and weekend, shift schedule, break times, day and night time, temperature, and to which extent sustaining processes (e.g., HVAC, CA) are operated outside regular production hours might, to different degrees, influence a factory's demand pattern. Variability and demand cycles can be within (sub-)second intervals/periods, e.g., if equipment is switched on, up to daily, weekly and seasonal patterns and cycles.

2.1.4 Energy Flexibility of Manufacturing Systems and Embodied Energy Storage

Influencing energy demand of manufacturing systems depends on the ability of processes and systems to adjust electricity demand according to control signals. The terms manufacturing (system) flexibility and, more specifically, energy flexibility are used in this context. Further, embodied energy storage in products is a closely related topic. All three terms are briefly discussed in the following.

Energy Flexibility of Manufacturing Systems

Flexibility of manufacturing systems can be defined in various ways. An early, comprehensive summary of available definitions and understanding of manufacturing flexibility has been provided by Beach et al. (2000). They refer to Gupta and Goyal, who suggested an initial definition of flexibility in manufacturing (systems) in 1989 (Gupta and Goyal 1989). Their definition, referring to earlier work from Buzacott and Mandelbaum (1985) and Mascarenhas (1981), is: "Flexibility is defined as the ability of a manufacturing system to cope with changing circumstances or instability caused by the environment" (Gupta and Goyal 1989, p. 120). Additional definitions have been proposed later on (see, e.g., Beach et al. 2000), which reflects the breadth of the subject. Another example is an early review of manufacturing flexibility, including its integration into manufacturing objectives and measuring flexibility, from Slack (1983). Nyhuis et al. further differentiate between flexible and changeable production systems. They note that a flexible production system is capable of changing parameters (e.g., production rate) within boundaries with reasonable effort and no additional investment. In contrast, a *changeable production system* is a system which can adjust parameters' boundaries with little effort (investment), relocating the system's flexibility boundaries (Nyhuis et al. 2009).

In the context of manufacturing system flexibility, *energy flexibility* refers to the potential of manufacturing systems to change energy demand. Few approaches exist to formally define and evaluate energy flexibility of manufacturing systems. A formal definition has been provided by Graßl et al., who define energy flexibility "as the ability of a production system to adapt itself fast and without remarkable costs to changes in energy markets" (Graßl et al. 2013, p. 303). A mathematical approach to define energy flexibility based on production theory is given in Kabelitz and Streckfuß (2014). The authors propose to define energy flexibility as the curvature of a system's energy demand function, i.e., the slope of the power demand function. Essentially, if an energy demand function is known, the first derivation over time results in the power demand function, and the second derivation over time (and thus the first derivation of the power demand function) provides the energy flexibility function (or parameter). However, the approach is mainly based on mathematical principles to derive a descriptive energy flexibility indicator and lacks consideration of technical details, especially in the context of manufacturing system (energy and material flow demand) dynamics. Further, no detail on how to influence energy flexibility or how to use the proposed indicator is given.

An illustrative example for energy flexibility in the context of (decentralized) VRE integration is given in Fig. 2.7 (generic example for visualization, see also Reinhart and Schultz 2014; Popp et al. 2013, and Liebl et al. 2015 for similar illustrations, also related to volatile electricity prices). Electricity demand from, e.g., a manufacturing system or machine and volatile supply, e.g., from on-site wind generation is illustrated in the upper left graph (a1). Three different periods (1, 2, 3) are distinguished, with 15 min duration each. In the initial setup (a1), demand is initially low (period one, e.g., during setup of a process), followed by higher demand (period two, e.g., due to processing a part) and subsequently reduced (between period two and three, e.g., due to switching products) and increased again after short time to higher (processing) demand in period three. Electricity supply is variable and exhibits short-term volatility and is generally higher in period one and three than in period two. The two graphs labeled energy flexibility I (b1) and II (c1) show different load shifting/energy flexibility examples where the sequence of the three initial demand periods has been changed. The lower graphs (a2, b2, c2) from Fig. 2.7 show grid supply and feed-in as the difference between supply and demand. In the initial setup (a2), total delta between supply and demand (and thus grid electricity flows) is characterized by an initial feed-in period (period one), followed by grid supply during period two and grid feed-in and then supply during period three. Using energy flexibility on the demand side, the middle graphs (b1, b2) show a demand time series for a case where the series' periods one and two are switched. Within the first 30 min, a significantly improved match between supply and demand is achieved and grid electricity flows are nearly halved compared to the initial setup (a2 vs. b2). However, if also period two and three are switched (energy flexibility II, c1 and c2 in Fig. 2.7), demand and supply fit is further improved and total electricity flows are approx. 10 % lower than in the energy flexibility I case (c2 vs. b2) and more than halved compared to the initial setup (c2 vs. a2). However, several demand and feed-in peaks are still present. In addition, and most importantly, feasibility of proposed load shifting requires sufficient energy flexibility of underlying processes: if the initially low demand (period one in the initial setup a1) was caused by process setup, periods two and three have to occur *after* setup and thus after the demand pattern from period one. Extending this logic, if the two high-level demand patterns are caused by a mandatory production sequence (initial setup a1, process from period two is required to finish before processes in period three can start), shifting periods two and three is also not feasible. In summary, energy flexibility enables VRE integration, but is subject to several (operational, technical) constraints. Other examples for applying energy flexibility methods include matching demand with volatile electricity prices (not further discussed here).

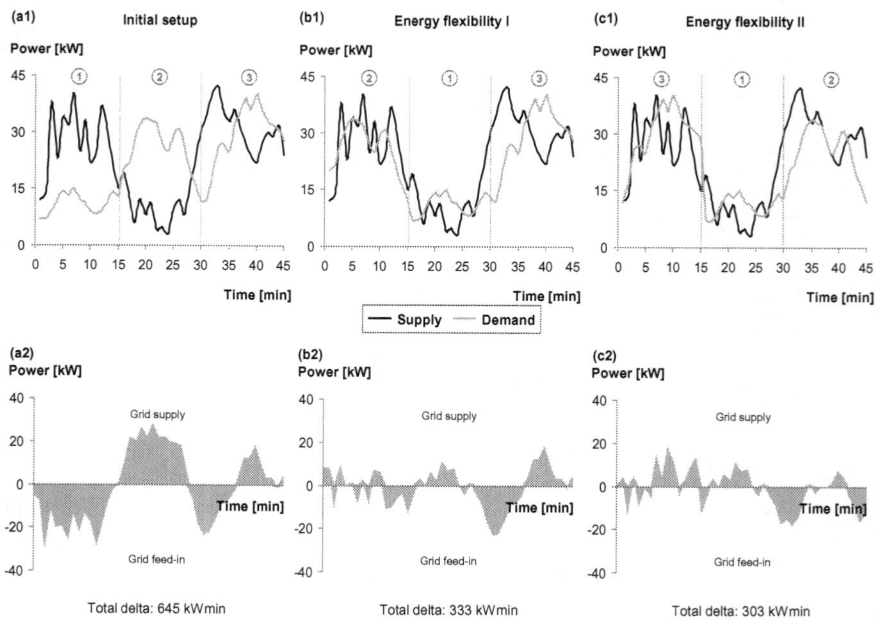

Fig. 2.7 Load shifting example for energy flexible manufacturing system (illustrative example, own illustration)

Embodied Energy Storage

Energy flexibility is used to enable and often related to *embodied energy storage* in products.[3] Assuming that processing a product requires a certain amount of energy, this energy can be allocated as embodied energy to the product. Embodied energy can be directly stored (e.g., in chemical bounds as a result of an (endothermic) reaction, e.g., in chemical process industries) or indirectly as a result of inefficiencies and thus dissipated energy into the environment of the product (not all energy required for processing is contained within the product). For example, milling and drilling of a workpiece and consequently joining with another workpiece requires a given amount of energy (see Sect. 2.1.3 for a discussion of manufacturing process energy demand). While most of the energy is not directly contained within the product but dissipated into the environment, the total amount of energy required can still be regarded as the embodied energy of the product. For further detail on embodied energy (and emissions) and embodied energy estimates for a set of materials see, e.g., Allwood and Cullen (2012) and Allwood et al. (2011).

Although not specifically referring to energy flexibility, the potential for storing electricity as embodied energy in products was already acknowledged in 1989 by Daryanian et al., who mention the option of "Rescheduling of electricity usage to the periods of lower costs, without curtailment of services. This becomes possible if the end products of electricity usage can be stored for later use" (Daryanian et al. 1989, p. 897) and further "[..] electrical energy can often be stored quite cheaply for a few hours or days on a decentralized basis, once it is embodied into intermediate or final products" (Daryanian et al. 1989, p. 899). More than two decades later, Lorenz et al. introduce the term 'energy efficiency 2.0' to include volatility of supply into the field of industrial energy efficiency (Lorenz et al. 2012). They explicitly note that industry can be used to store energy from volatile supply and that factories should be, in the context of energy efficiency 2.0, become active energy/power market participants instead of passive consumers, enabled by energy flexibility. In general, embodied energy storage can be used to store a specific type of energy (e.g., electricity from VRE supply) by demanding this type of energy during processing of a product. Consequently, the amount of embodied energy which can be stored depends on the ability to store (intermediate) products. Further, selectively increasing or decreasing energy demand to increase demand of a certain energy supply type strongly depends on the manufacturing system's ability to adjust energy demand, and thus its energy flexibility. Therefore, previously described constrains (e.g., material flow constraints, shifting and switching of processing sequence) need to be taken into account in the context of embodied energy storage.

[3]For further detail on embodied energy storage as a mean to integrate VRE see Beier et al. (2016), on which this section is partly based on.

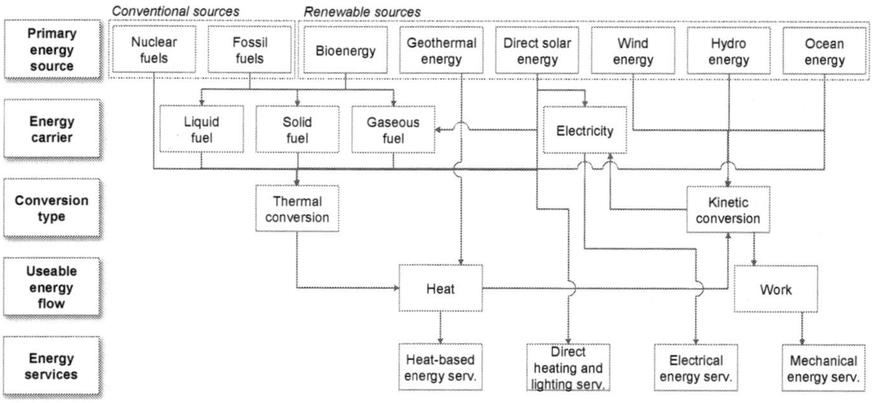

Fig. 2.8 Conversion pathways from energy source to service (original from IPCC 2011, p. 38, own illustration)

2.2 Conventional and Renewable Electricity Supply

After focusing on the electricity demand side and the case for manufacturing systems in particular, this section provides background on electricity supply. Definitions are given first, followed by detail on conventional, renewable, and decentralized electricity generation. The interaction between electricity demand and supply is then described (distribution, markets, grid stability), and VRE integration into an electricity grid discussed.[4]

2.2.1 *Definition and Classification*

Energy is defined as the capability of a system to generate an outside effect, i.e., to perform work (Quaschning 2011). Different primary energy sources are used to provide various energy services, enabled by energy carriers, conversion, and energy flows as illustrated in Fig. 2.8.[5] An *energy service* can be understood as "the tasks performed using energy" (IPCC 2011, p. 40), and includes, for example, mobility, space comfort, communication, lighting, or manufacturing and assembly of a product.

[4]Note that energy supply and generation are generally used interchangeably within the following. However, *generation* is commonly used when describing the (technical) conversion, while *supply* is used in a broader meaning which might include a market-related or energy distribution-related context.

[5]This visualization is one possible classification scheme, while other illustrations/classifications might be possible. However, this overview is chosen as it highlights the multitude of different conversion options from energy source to energy service.

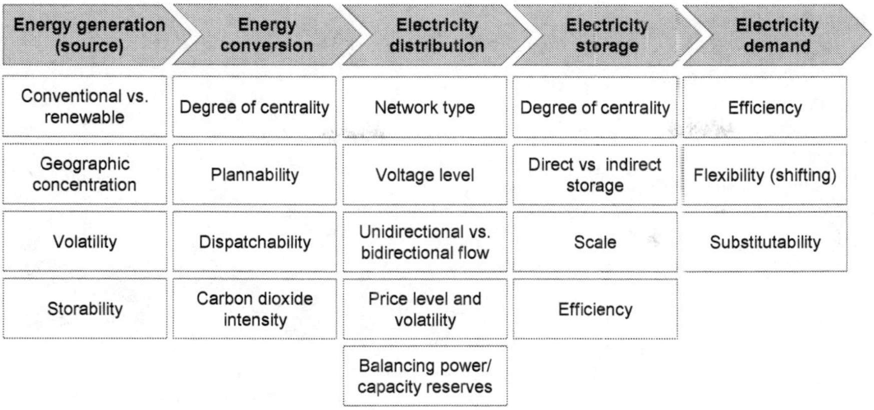

Fig. 2.9 Overview and structure energy generation to electricity demand (own illustration)

Electricity generation is commonly classified according to primary energy sources and split into nuclear energy, fossil fuels, and renewable energy (IPCC 2014; Schwab 2015). Further, the terms conventional and renewable electricity generation are used to differentiate energy sources (Kosmadakis et al. 2013). *Conventional electricity* generation entails all large-scale/commercial electricity generation from non-renewable sources (fossil fuels and nuclear energy), while *renewable electricity generation* refers to electricity generation from renewable energy sources.

Consistent with the IPCC's classification, RE resources include (photosynthetic) biomass energy, geothermal energy, direct solar energy, wind energy, hydropower and ocean energy (tide, waves and thermal energy) (IPCC 2011, 2014). Some of these RE forms are exhaustible (e.g., geothermal, biomass) and indirectly rely on solar radiation (i.e., hydropower, wind) (IPCC 2014; Ellabban et al. 2014). Within this context, *renewable electricity* is understood as electricity used as energy carrier for energy from primary renewable energy sources (compare also to Fig. 2.8).

Figure 2.9 gives an overview of different topics and characteristics which need consideration when discussing energy generation to electricity demand with a focus on (variable) renewable energy sources. As such, the structure in Fig. 2.9 highlights the different topics which are addressed in the remainder of this chapter. For energy generation, a focus is set on conventional and renewable generation and their characteristics (e.g., volatility). In this context, different energy conversion methods are discussed. Electricity distribution focuses on different network types and technology, as well as the influence of integrating VRE on electricity markets and grid stability. Electricity storage is discussed as a mean to match supply and demand (timely differences), e.g., through indirect/centralized (e.g., pumped hydro) and direct/decentralized (e.g., battery) storage. The last area, electricity demand, has already been discussed from a demand side perspective in the previous section, and therefore supply-side options for flexibilizing electricity demand are summarized.

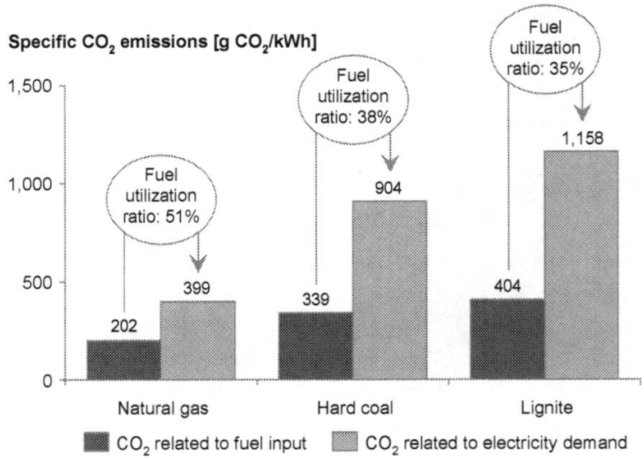

Fig. 2.10 CO_2 intensity from natural gas, hard coal, and lignite based electricity generation: related to fuel input and final electricity demand, percentages indicate fuel utilization ratio (data from Icha 2015, own illustration)

2.2.2 Conventional Electricity Generation

As outlined before, conventional electricity generation includes fossil fuel based generation and nuclear-based generation. Most important energy carriers within these categories include hard coal, lignite, mineral oils, natural gas, and nuclear fuels (e.g., uranium, thorium) (Schwab 2015). Large-scale conversion of these carriers is performed via thermal conversion (combustion or nuclear fission[6]) to heat, which is converted in turbines to kinetic energy and subsequently to electricity in generators (IPCC 2011; Schwab 2015, cf. Fig. 2.8). Common power plant types include thermal power plants (i.e., steam-driven, including, e.g., coal-fired and nuclear fission), gas power plants (natural gas) or combined cycle plants (gas power plant with attached steam power pant). Several combined and advanced technologies have been and are developed to increase efficiency of conventional power plants, e.g., cogeneration of heat and power to supply customers (e.g., industry) with both electricity and heat (e.g., steam).

As noted in the introductory chapter, a global and national target is set towards reducing GHG emissions, among others from energy and power supply. As such, a focus within this section is set on the relevance of conventional electricity generation for contributing to and reduction of GHG emissions rather than discussing technical details of conventional generation (for a technical overview and additional details on electricity generation in power plants see, e.g., Schwab 2015).

Figure 2.10 illustrates specific CO_2 emissions from natural gas, hard coal, and lignite. While natural gas exhibits lowest emissions related to fuel input, the difference between all three energy carriers for final demand is not only caused by higher input-related emissions but amplified due to worse fuel utilization ratios for hard coal and lignite (i.e., considering conversion and other inefficiencies from fuel input to final electricity demand) (Icha 2015).

[6]Nuclear fusion is not included here as not commercially mature and not used on a large-scale.

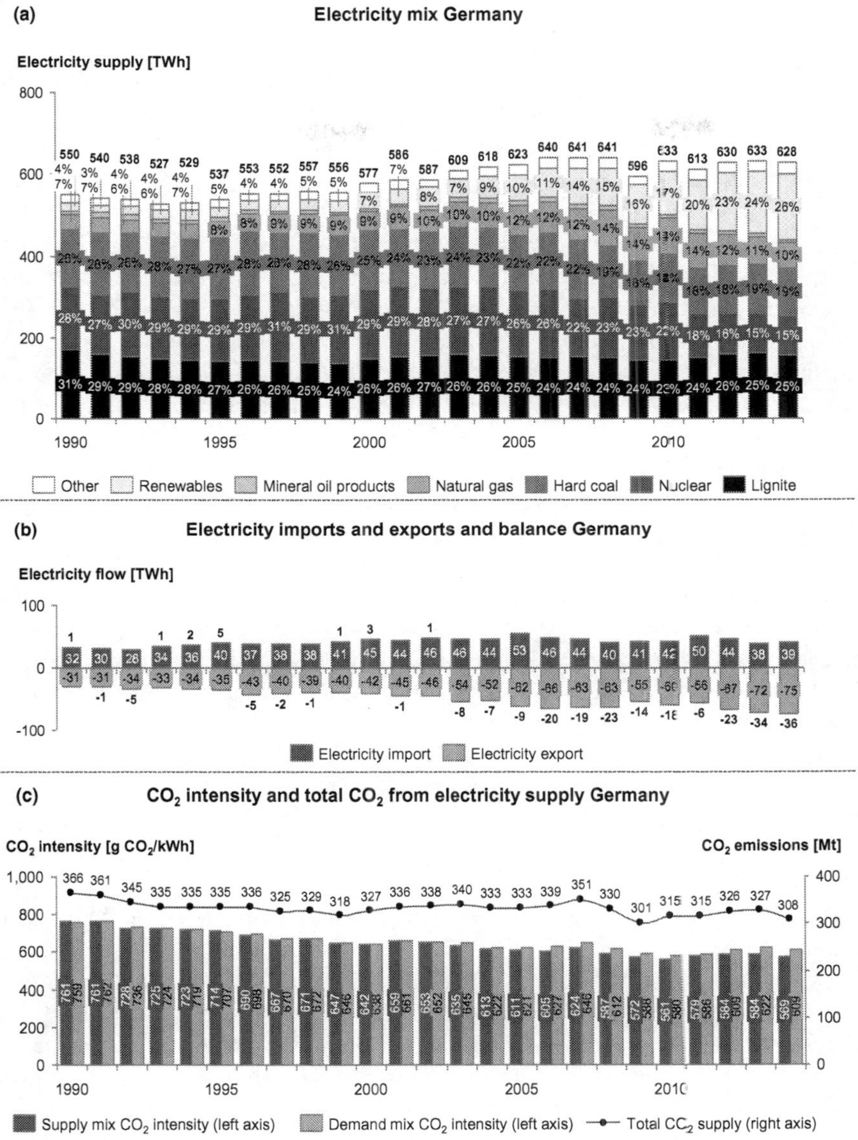

Fig. 2.11 German electricity mix (**a**), electricity imports, exports and balance (**b**) and CO_2 intensity and total CO_2 emissions (**c**) from 1990 to 2014 (data from Icha 2015 and AG Energiebilanzen e.V. 2015, own calculations and illustration)

Different specific CO_2 emissions are especially relevant when electricity supply mix changes, for example, as a result of integrating larger shares of RE generation. Utilizing Germany as an example, Fig. 2.11a illustrates the national energy supply mix from 1990 to 2015. The German energy mix is characterized by a fairly stable lignite supply share of approx. 25 %, reduced nuclear and hard coal supply shares within the last years, slightly increasing natural gas generation and significant increase in renewable energy sources as a result of the *Energiewende*. Total electricity supply has been stable within the last ten years and is slightly higher compared to 1990–2000 levels. Further, considering the shift in supply source shares, renewable generation is mainly substituting nuclear and hard coal generation (partly due to a politically driven nuclear exit and newly built and dispatched lignite capacities). Considering electricity imports and exports and thus net demand balance (b in Fig. 2.11, total electricity demand equals supply (a) plus import–export balance (b)), total electricity demand has been decreasing from an all-time high of 621.5 TWh in 2007 to 592.2 TWh in 2014. However, recent years' import and export balances indicate that a substantial share of domestically generated electricity is exported (approx. 12 % of total generation), while also importing approx. 6 % of total generation, resulting in a net differential of approx. 6 %. This indicates an increasing net oversupply, while national supply and demand are more frequently dynamically unbalanced (high imports *and* exports). For example, excessive RE supply is exported into the connected European network, while supply gaps are filled by importing electricity.

Both the shift in energy sources and increasing net negative trade balances are reflected in CO_2 emission factors for German electricity supply and demand. Figure 2.11c illustrates specific CO_2 emissions per kWh both for electricity supply and demand and total emissions from electricity supply. Comparing values from 1990–2000 in relation to most recent years, all indicators (specific and total) are reduced. However, while specific supply and demand mix emissions were relatively similar until 2002, a shift has occurred after 2002, towards higher specific emissions in the demand mix compared to the supply mix. This indicates that, on a relative basis, carbon-intensive generation was demanded domestically, while low-carbon electricity has been exported. An increased share of renewable energy generation (a in Fig. 2.11) achieved to lower specific supply emissions, however, increased exports (b in Fig. 2.11) of low-carbon intensive electricity result in relatively higher demand mix emissions.

Combining specific CO_2 emission values from Fig. 2.10 with relative supply share changes from Fig. 2.11a, it becomes evident that even though renewable share increased (which is accounted for as zero CO_2-emitting), reducing low(er)-carbon generation from hard coal and nuclear results in relatively stable, and periodically (2010–2014) increasing total emissions and demand mix specific relative emissions (c in Fig. 2.11). This highlights the requirement to not only deploy additional renewable supply sources, but to successfully integrate these into an existing supply and demand system: if possible, low(er) carbon intensive sources should not be substituted by high(er)-carbon intensive sources (e.g., as a result of profitability changes). Further, measures have to be found to increase (domestic) demand of renewable generation rather than exporting surplus generation to effectively reduce relative (per kWh) and absolute carbon emissions from electricity supply *and* demand, which directly translates to the research question formulated in Chap. 1.

2.2.3 Characteristics of Renewable Electricity Generation

Renewable energy is understood as energy from sources, which, in the sense of renewable resources, natural processes replenish at rates relevant in relation to their use (Perman et al. 2003; IPCC 2011; Ellabban et al. 2014). Renewable resources can be divided into *stock* and *flow* resources, i.e., resources which have a certain stock amount which is replenished (e.g., biomass) and resources which have a (constant) inflow (e.g., sun) (Perman et al. 2003). As such, renewable stock resources can be extinct, for example, if a biologically-based resource (biomass) is utilized at rates which prevent the resource to be replenished at a sufficiently high enough rate to replace the utilized amount. In contrast, flow resources are considered as non-exhaustible, at least within their current utilization patterns (e.g., wind) (Perman et al. 2003; IPCC 2011). Non-renewable (energy) resources include resources which are replenished over time (i.e., coal through conversion of biomass), but replenishment rates are negligible compared to utilization rates (IPCC 2011).

Certain RE sources are *variable renewable energy* (VRE) sources, which are usually non-dispatchable, i.e., power generation cannot be increased as desired (unless output was previously curtailed), and only be reduced (which results in opportunity losses of generation).[7] This characteristic substantially differentiates VRE from fossil fuel based plants, from which output can be (within certain ranges) controlled. Further, output from VRE is not only non-dispatchable, but also significantly more volatile than from dispatchable fossil fuel plants. Figure 2.12 shows the volatility of wind and solar generation. Illustrated are wind and solar generation time series from solar panels and a wind turbine installed at the Institute of Machine Tools and Production Technology (IWF), TU Braunschweig, Germany, for five days in September. Both time series are indexed to their maximum value to facilitate comparison. Referring to Fig. 2.12, the following observations with regards to variable generation and related challenges can be made:

- Short-term (minute) fluctuations can be relatively large, especially for wind generation.
- Wind generation is characterized by short-term (minute) variability, but also exhibits longer term variability (i.e., hour to daily generation-level changes).
- Solar generation can be very stable (September 6, likely a very sunny day) but can also be strongly reduced and volatile (e.g., September 7, sunny with few clouds and September 8 to 10 with short-term sunny periods during a cloudy day).
- Solar generation is, for long periods (night), zero. While this is a trivial fact, it is still relevant in combination with wind generation: wind can occur at any time of

[7] The term variable/variability and volatile/volatility are both used here to describe a set of datapoints that are not equal, i.e., distributed. While variability might usually refer to a general set of datapoints, different options or characteristic, volatility is more commonly used in the context of time series data (and is, as such, also encompassed by the term variability). For example, in finance, volatility of a time series usually refers to the standard deviation over a given period. The terms variability and volatility are both used here, while variability is regarded as a broader term (including volatility) while volatility is only used in the context of (stochastic) time series.

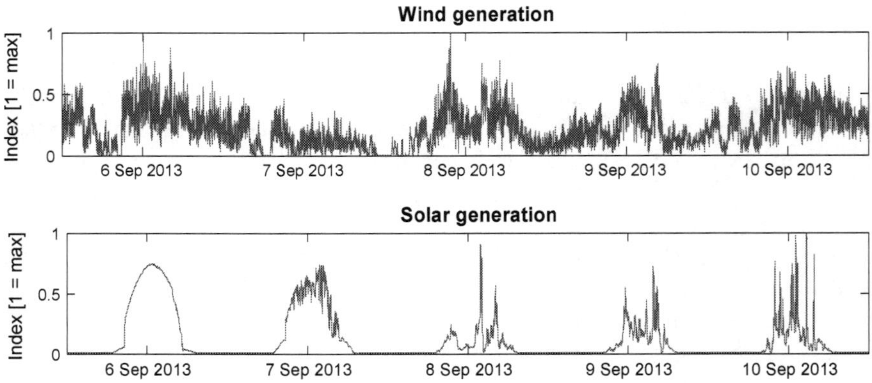

Fig. 2.12 Wind and solar generation at IWF, TU Braunschweig (latitude 52.2767, longitude 10.5369) from September 6 to 10, 2013, minute resolution, normalized to sample period's maximum generation (own data and illustration)

the day, allowing for electricity supply during night. However, as wind is a rather unstable source, high wind generation can occur during periods of high solar output (September 6) and low solar output (night from September 6 to 7), but low wind generation can also occur during periods of high solar output (September 7) and low solar output (night from September 7 to 8). As a result, combined generation can be favorable, but can also increase variability.

Table 2.1 provides a qualitative assessment of renewable energy sources (classification according to IPCC) and fossil fuel energy against their individual volatility, storability, and dispatchability. Although the assessment can only provide an indication, it becomes evident that especially wind and solar energy are unfavorable in several dimensions compared to biomass energy and hydropower, and even more unfavorable compared to (traditional) fossil fuel based generation. This comparison further highlights the need for new methods to compensate disadvantageous characteristics such as high volatility or missing dispatchability of renewable energy sources.

Further details on common RE technologies can be found in Geitmann (2010), a comprehensive overview of RE to mitigate climate change, including extensive data, is provided in IPCC (2011). Predicting electricity generation from VRE is a central concern to increase planning reliability of electricity supply and demand. For a comprehensive overview of existing research and methods for predicting solar output using artificial neural networks see, e.g., Qazi et al. (2015). Page 182 in IPCC (2011) includes an overview of 46 RE technologies and their technical maturity. In addition, 23 of these technologies are classified as decentralized (on-site) technologies, highlighting the relevance of decentralized RE technologies, which are discussed in the next section.

Table 2.1 Comparison of renewable energy sources and fossil fuel energy (very favorable (+ +), favorable (+), neutral (o), unfavorable (−), very unfavorable (− −); own assessment based on references from this section)

	Volatility	Storability	Dispatchability
Solar energy	−	−	− −
Biomass energy	+	+ +	+
Wind energy	− −	− −	− −
Ocean energy	o	− −	− −
Hydropower	+	+	+ +
Geothermal energy	+ +	o	o
Fossil fuel energy	+ +	+ +	+ +

2.2.4 *Decentralized Generation*

Decentralized renewable energy generation is "deployed at the point of use (decentralized) in rural and urban environments, whereas others [RE technologies] are primarily deployed within large (centralized) energy networks" (IPCC 2011, p. 7). RE sources play a vital role in decentralized generation. Decentralized renewable generation capacity has been increasing in recent years, technologies include small hydropower, photovoltaics (PV), wind generation, and biomass (IPCC 2014). Several RE sources are inherently geographically spread out, i.e., wind resources and solar radiation are not concentrated within a single, local point as, for example, fossil fuel based generation facilities (where concentrating fossil energy has been performed by transportation of fuels to a power plant). RE is commonly available within a larger geographic region, i.e., coastal areas for wind and elevated areas for PV (although some preferred local spots exist where to deploy generation capacities, e.g., hills for wind turbines).

Decentralized generation is not unique to a specific primary RE source. For example, wind-based generation can be deployed on a centralized basis using large turbines which commonly require grid integration (especially offshore), while small turbines or wind kites are used for decentralized generation (IPCC 2011). For the case of solar-based generation, PV electricity generation can easily be scaled by installing a desired set of panels (given enough space), and thus PV electricity generation can be deployed centralized (e.g., PV plants) or decentralized (e.g., on rooftops). On the other hand, low-temperature solar thermal energy is mainly used in decentralized applications as a consequence of low conversion and transportation efficiency. However, considering given examples and above definition, decentralized generation can include PV generation on household rooftops with direct use in a single house or, for example, wind generation in an urban environment and use within a local distribution grid.

Most industrial, and especially production and manufacturing companies, have extensive on-site (decentralized) electricity generation. In 2013, 8.9 % or 45 TWh of total gross German electricity generation was from industry, which equates to approximately 18 % of total industrial electricity demand (DESTATIS 2014, 2015, own calculations). The Association of German Chambers of Commerce and Industry (DIHK) estimates that approximately 25,000 companies in Germany produce and utilize their own electricity, with another 25,000 companies planning to set up own supply. In summary, 16 % of all firms enacted measures for own supply in 2013. Solar generation had the largest share of all utilized technologies (62 % realized or planned) (DIHK and VEA 2014). Solar- and wind-based generation have an inherent decentralized structure and are therefore of special interest for decentralized or industrial on-site generation. As a consequence, with an increase in VRE generation, decentralized generation is expected to increase as well (DIHK and VEA 2014). Further, the DIHK notes that for most companies investments into generation capacities are most useful when maximizing utilizing of self-generated electricity (DIHK and VEA 2014).

2.2.5 *Electricity Distribution, Markets, and Grid Stability*

After highlighting the different energy and electricity supply options, a brief overview of electricity distribution and the supply–demand connecting element, electricity markets, is given, followed by emerging concerns in relation to VRE integration.

Electricity grids are required to connect electricity supply sources (i.e., power plants) with electricity customers (i.e., industry). As such, different grid types exist. On the highest (voltage and regional) level, extra high voltage and high voltage networks (380 and 220 kV) fulfill transmission tasks and are therefore part of a transmission network (Schwab 2015). In Europe, 41 transmission system operators (TSOs) from 34 countries are combined in an European system, the European Network of Transmission System Operators for Electricity (ENTSO-E) (ENTSO-E 2015). However, not all member TSOs are also part of a technically synchronized grid, which is limited to an area formerly known as the Union for the Co-ordination of Transmission of Electricity (UCTE), which included 29 TSOs from 24 continental European countries, and which supplies approx. 500 million people through a synchronized network (Schwab 2015; ENTSO-E 2015).

Aside from transmission networks, distribution networks exist to connect single customers with transmission networks. They are normally sized as high (110 kV), medium, and low voltage networks. Table 2.2 summarizes different network typologies and their characteristic voltages. A main difference between transmission and distribution networks are their main energy flow directions (in Germany/Europe): while transmission networks are frequently designed as a mesh structure where different energy flow directions are feasible, distribution networks usually exhibit a unidirectional flow structure (from transmission network to customer) (Schwab 2015). However, with an increase in decentralized (renewable) electricity generation, distribution networks are subject to an increasing amount of bidirectional energy flows, causing technical difficulties as, for example, security systems have

Table 2.2 Nominal and maximum power ratings of different electric power networks (from Schwab 2015, p. 36, translated from German and resorted)

Network type	Nominal rating (U_n)	Maximum rating (U_m)
Extra high and high voltage networks		
Transmission networks	380 kV	420 kV
Transmission networks	220 kV	245 kV
High voltage networks	110 kV	123 kV
Medium voltage networks		
Special customers	35 kV	40.5 kV
City distribution networks	20 kV	24 kV
City distribution networks	10 kV	12 kV
Industry and power plant own demand	6 kV	7.2 kV
Low voltage networks		
Large industry, own demand	400 / 690 V	
Domestic, commercial, industrial	230 / 400 V	

not been designed for this purpose. Consequently, direct/local demand of decentralized generated electricity can reduce requirements for technical and network layout adjustment measures.

Electricity markets connect electricity supply and demand via price. Depending on time horizon until fulfillment of contracts, markets are commonly separated into capacity and forward markets (months to years until delivery), day-ahead markets (commonly next day) and intraday/real-time markets (quarter hour to hours) (U.S. Department of Energy 2006; Schwab 2015). Within recent years, energy exchanges have more and more replaced direct (over-the-counter, OTC) contracting. For Europe, the European Energy Exchange (EEX) is one of the leading energy exchanges (EEX 2015; Schwab 2015).

For renewable electricity integration, the Merit Order Effect is of special interest in a supply-demand-price context. Commonly, electricity generation plants are ranked according to their marginal (variable) operating costs, resulting in an increasing step function of generation cost over capacity (merit order) (U.S. Department of Energy 2006; Schwab 2015; Sensfuß et al. 2008). Assuming a relatively (short-term) inelastic electricity demand, the power generator which marginally fulfills power demand sets the market price. If renewable energy sources with lower marginal costs than current market price or mandatory feed-in ahead of other generation capacities (e.g., in Germany) are introduced into the market, the marginal generator might be pushed out of the market and wholesale price is reduced. For the case of fluctuating RE feed-in, market prices can become more volatile and dependent on (relative) VRE supply share. Figure 2.13 illustrates the Merit Order Effect. The figure shows the hourly electricity price (Phelix) over wind and solar generation share of total load in Germany for 2013 (8,760 data points). Although the statistical relationship is not very strong (coefficient of determination is 0.2154, different total system load levels and

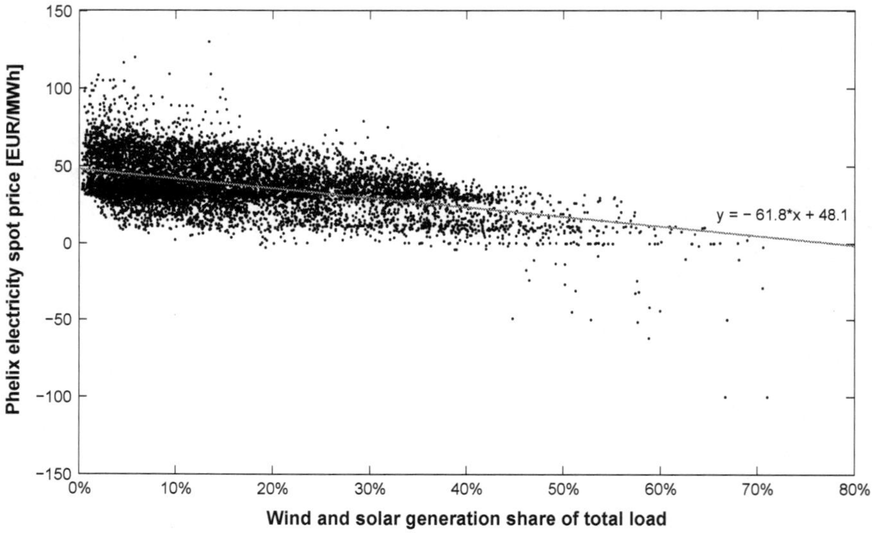

Fig. 2.13 Illustration of the Merit Order Effect: hourly spot electricity price over renewable generation load share for Germany in 2013 (own calculation and illustration, data from EPEX SPOT 2014; EEX 2014a, b; ENTSO-E 2014)

other market determining factors are also influencing the price), some observations can be made:

- Electricity price is, on average, negatively correlated with wind and solar share of total grid load.
- Extremely high prices generally occur when wind and solar share is low.
- Prices are never extremely high when wind and solar share is high.
- Negative prices are never realized when wind and solar share is low.
- Negative prices occur only when wind and solar share is high.

In summary, the Merit Order Effect has multiple effects on price as a result of integrating VRE into larger power grids and markets, and influences need to be considered and compensated by utilities and TSOs (e.g., peaking-power plants and reserve capacities might become uneconomic if pushed out of the market too often/if they are only required for a very limited number of hours per year). For more detail on the Merit Order Effect and the Merit Order in general see, e.g., Sensfuß et al. (2008) or Bundesnetzagentur (2014).

Considering time-to-delivery to differentiate electricity market contracts, the shortest time-to-delivery periods are ancillary service markets, which provide balancing power to ensure a stable power grid (grid frequency) (U.S. Department of Energy 2006; ENTSO-E 2009). Balancing power is activated to either increase or decrease supplied power to stabilize the grid frequency, i.e., power is reduced if grid frequency increases (e.g., as a result of a demand decrease or if another power supplier suddenly increases power) or power supply is increased if grid frequency decreases.

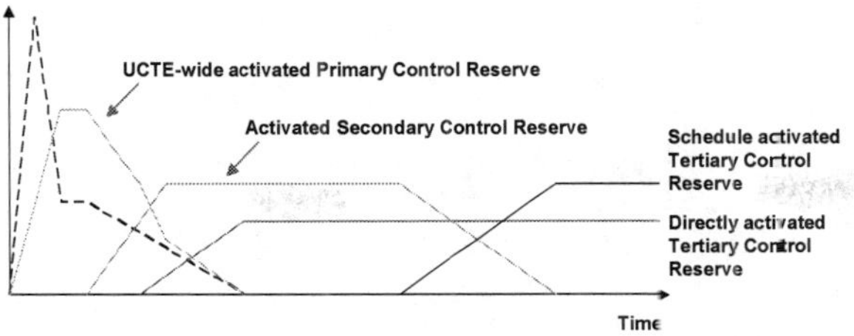

Fig. 2.14 Activation of frequency regulation as a result of a grid frequency change (*dashed line*) (from ENTSO-E 2009, p. P1-3, own illustration)

Figure 2.14 illustrates the dispatch order of primary (within seconds), secondary (within few minutes), and tertiary reserve (within minutes) capacities as a response to a grid frequency change (dashed line). Integration of VRE into the power grid might require additional balancing capacities to accommodate sudden power increases and drops due to stochastic wind and solar generation output (Bundesministerium für Wirtschaft und Energie (BMWi) 2013).

2.3 Integrating Renewable Electricity Supply

Several options exist to integrate (variable) renewable electricity supply into existing power systems. Energy storage (large-/small-scale) and influencing demand to better match supply (DSM) are common options to integrate fluctuating supply. In the context of influencing demand, energy flexibility of manufacturing systems has been discussed in Sect. 2.1.4. Therefore, energy storage and DSM as additional integration options and their sub-categories are discussed in the following (for an overview see Fig. 2.15, note that the dashed line indicates that integration options might be related and/or complement each other).[8]

2.3.1 Energy Storage for Renewable Energy Integration

Options discussed here for energy storage to integrate renewable energy include grid-wide, large-scale options (e.g., pumped hydro) and technologies which can be deployed decentralized and thus on a smaller scale, e.g., to integrate decentralized RE generation for direct (on-site) demand.

Centralized, Large-Scale Energy Storage

Since the widespread deployment of large-scale power plants which have limited options to change (reduce, increase) output within shorter (minute to hour) time

[8] This section is partly based on Beier et al. (2016).

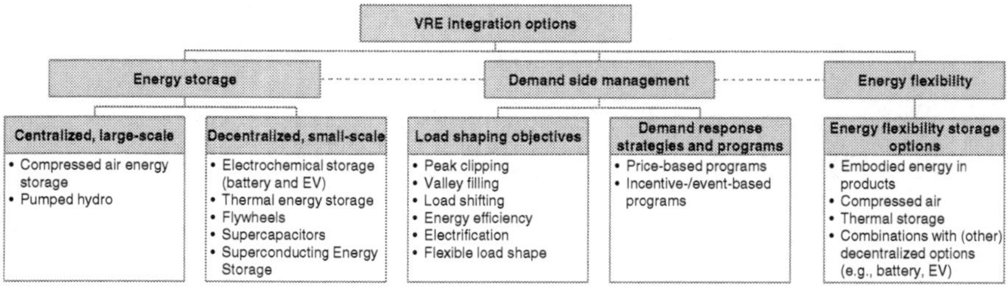

Fig. 2.15 Overview of VRE integration options (own illustration)

frames, grid-wide electricity storage has been used to overcome temporal mismatches between a relatively inelastic demand and an also inelastic base-load power plant fleet (cf. also Fig. 1.4 - electricity demand in Germany is rarely below 40 GW during night on weekdays, but as low as 30 GW on Saturday nights, while demand peaks can be above 60 GW). Power plants with low marginal generation costs (low position on Merit Order Curve, e.g., lignite and nuclear) exhibit times of oversupply, e.g., at night, while reducing output is either technically difficult or uneconomic (e.g., due to set efficient operating range or high fixed costs). As such, options for (short-term) energy storage have been developed, among them compressed air energy storage (CAES) and pumped hydro storage, which are the two main large-scale available technologies (IPCC 2011).

CAES technology is based on storing energy in compressed air. Available surplus electricity (e.g., during periods of low prices, which indicate supply surplus) is compressed in a turbine and stored in a large-scale containment, for example, an underground (natural) cavern. If electricity is required (if prices are higher), high-pressure air is released through the turbine and electricity is generated. However, as the air is cooling down as a result of its expansion, heat is required during the process, which is, for example, provided by additional natural gas firing or from heat exchangers which can store surplus heat from compression. Consequently, round-trip efficiency for CAES is relatively low and within the range of 50 %. Currently operated plants can be found in McIntosh, Alabama, USA (110 MW, commissioned 1991) and in Huntorf, Germany (290 MW, from 1978) (Crotogino et al. 2001). Low conversion efficiency of CAES provides economic challenges as profitable operation strongly depends on electricity price volatility. Further, ecological challenges and additional environmental impact is caused by plant setup and operation (compared to direct electricity demand). Therefore, several new CAES-based technologies are subject of research and development efforts. Among them is the isobaric adiabatic compressed air energy storage plant with combined cycle (ISACOAST-CC) with a targeted 80 % efficiency and the advanced-adiabatic compressed air energy storage technology (AA-CAES) with 70 % target (Bullough et al. 2004; Zunft et al. 2006; Nielsen and Leithner 2009; Karellas and Tzouganatos 2014).

Another large-scale energy storage technology, pumped hydro storage, is a mature technology for grid-wide electricity storage. Water is pumped from a lower-level reservoir (e.g., a pond or river) into a higher-level storage basin. The process is reversed for electricity generation, energy stored as potential energy is converted back to electricity. Energy efficiency varies between 70 and 80 % in practice (Rehman et al. 2015).

Both CAES and pumped hydro have the additional advantage of relatively fast response times and can thus be (depending on specific technical realization) also used to provide ancillary services, such as tertiary reserves. Their fast response times make especially CAES plants interesting for VRE integration (Lund and Salgi 2009; Lund et al. 2009). Pumped hydro storage has relatively high conversion efficiencies and low marginal costs, making it the only currently economic viable large-scale storage solution (Deutsche Energie-Agentur GmbH (dena) 2010). However, installing pumped hydro not only depends on the presence of suitable geological conditions, but might also be subject to ecological and political resistance, even if options for additional storage would exist (Germany Trade and Invest 2015). Consequently, capacities for additional storage is limited in several regions, for example, in Germany. In order to further increase electricity storage options in the Central European Grid, additional pumped hydro capacities exist in Scandinavian countries and the Alps. Norway alone is estimated to have another 10–25 GW of pumped hydro storage capacity, which is about equivalent to currently available capacity (IPCC 2011). Nonetheless, transportation of surplus electricity from Central to Northern Europe requires a significant increase in transmission capacities (e.g., sub-sea cables).

Several additional storage technologies are known and their suitability for large-scale energy storage investigated. Nonetheless, technical, economic, and environmental barriers exist, e.g., economic challenges for hydrogen (fuel-cell) conversion strategies/other power-to-gas methods, environmental impact challenges from electrochemical (battery) storage or high self-discharging rates of flywheels, which are a technical challenge (while most named challenges are related to each other) (Deutsche Energie-Agentur GmbH (dena) 2010).

Decentralized, Small-Scale Energy Storage

Decentralized energy storage is used to overcome temporal mismatches between on-site electricity demand and decentralized supply. Although decentralized generation could be fed into the grid and stored in previously discussed large-scale energy storage facilities, several advantages of decentralized storage exist. On a grid-wide scale, less infrastructure (transmission capacities) is required to transport decentralized generation to centralized storage and back to customers. Further, if decentralized storage achieves to flatten grid demand and feed-in peaks from on-site generation, less ancillary services (backup generation) within the power grid are required. From a local (company) point of view, direct demand of on-site generated electricity is favorable due to economic benefits: on-site capacities have usually lower marginal generation costs than external grid supply costs and network charges and taxes can be (partly) avoided, for example, on-site demand is subject to lower/no subsidy-tax under the *Erneuerbare-Energien-Gesetz* in Germany (DIHK and VEA 2014). Table 2.3 summarizes five decentralized energy storage technologies, including a brief description, and factors for application in a VRE integration context.

Table 2.3 Overview of energy storage technologies in the context of decentralized VRE integration (based on references from this section and from Beier et al. 2016)

Technology	General principle	Current application for VRE integration	Advantages	Disadvantages
Electrochemical storage (battery)	Chemical reactions to store and release energy	Short-term demand/supply matching (minutes to hours) and for ancillary services	• Input and output is electricity • Fast response • High energy storage feasible • Several different technologies and characteristics	• Economic challenges • Fast degradation (cycle lifetime) • Hazardous materials
Thermal energy storage	Temperature or latent (phase changing) heat storage	Direct heating/cooling (solar energy) or from electricity if economic (e.g., fuel switching)	• High capacity • Low losses • Direct use (heat/cool) possible • Ecological, save	• Round-trip (back to elec.) inefficient • Without conversion: direct use limits applicability • Sufficient material/mass for storage necessary
Flywheels	Rotating mass energy storage	(Very) short-term demand/supply matching, special applications (e.g., sensitive environment)	• Fast charging/discharging • Nearly no hazardous materials • Low degradation, wear-out and maintenance	• High investment required (economically challenging) • Low total energy storage • Significant energy losses over time
Supercapacitors	Electric field energy storage	Very short-term storage and smoothing of, e.g., power output and frequency control	• Fast charging/discharging • Low maintenance and long lifetime • (Direct) electricity storage • Relatively high efficiency	• Energy per volume low • Economic challenges
Superconducting Magnetic Energy Storage	Electromagnetic field energy storage	Useful for frequency regulation, power quality applications (e.g., uninterrupted supply)	• Fast charging/discharging • Electricity conversion efficient • Limited hazardous materials	• Energy per volume very low • Energy losses over time high • Economic challenges • Long-term efficiency due to cooling reduced

Electrochemical storage options (batteries) are a well-known technology for decentralized electricity storage. However, technical maturity is largely dependent on considered technology. In general, three different technology types are common (Deutsche Energie-Agentur GmbH (dena) 2010; Oberhofer 2012; Oertel 2008):

- Conventional batteries (e.g., lithium-ion/-air/-polymer, lead–acid, nickel–cadmium, zinc-air, sodium–sulfur batteries)
- Redox-flow batteries (e.g., zinc–bromine, vanadium redox)
- High temperature batteries (e.g., sodium–sulfur, sodium–nickel chloride).

Conventional batteries include technical mature types like lead-acid batteries, which are frequently used for decentralized autarkical VRE integration with backup generation (e.g., diesel) (Bernal-Agustín and Dufo-López 2009). Availability of more cost efficient options like pumped hydro has prevented large-scale deployment of lead-acid batteries in applications with grid access (Oberhofer 2012). Further, their relatively low energy density (volume and mass) limits applications, especially in mobile application, which has caused extensive research towards improvement of lithium-based batteries. Strong research efforts are improving technical, economic, and ecological factors of lithium-based batteries, especially for use in electric vehicles (EVs) (Gallagher and Nelson 2014). Future development, especially reduced cost, might allow utilizing lithium-based batteries for widespread VRE integration, while they are currently mainly used for balancing power (as opposed to large-scale energy storage) applications.

Redox-flow batteries are based on a similar technological principle than conventional batteries. However, their electrolyte is stored in attached tanks and can be changed during operation. Therefore, energy and power parameters can be configured separately, and energy capacity simply be increased by adding additional electrolyte holding capacity. Further, spent electrolyte might be replaced, making redox-flow batteries potentially suitable for mobile applications (Oertel 2008; Oberhofer 2012; Deutsche Energie-Agentur GmbH (dena) 2010; Denholm et al. 2010). An example for a scalable vanadium redox battery system is the so-called *CellCube* by Gildemeister energy solutions (part of DMG MORI), which is, among others, used for energy-autarkic systems and to smooth VRE output (Gildemeister energy solutions GmbH 2016).

High-temperature batteries need to be heated during operation (commonly above 300 °C) as their electrolyte is solid and electrodes are liquid (molten). Their reduced efficiency due to heating and hazardous operation is offset by a long lifespan and high energy density (Oberhofer 2012; Oertel 2008).

Considering batteries for VRE integration, main characterizing differences between battery types and technology are their power and energy configurability. Round-trip efficiency and response time are usually high, which is offset by potential use of hazardous (toxic/flammable) and environmentally unfriendly materials, combined with (currently) relatively high costs and limited lifetime (Deutsche Energie-Agentur GmbH (dena) 2010; Oberhofer 2012).

Thermal energy storage relies on the principle of using heat and cooling capacities of thermal mass to store energy, i.e., in building thermal mass, cool storage tanks (water and ice) or subsurface solid rock formations. Either temperature difference or latent energy storage (phase change of materials) can be applied (Arteconi et al. 2012; Oertel 2008). Different conversion pathways exist, for example, electricity-heat, electricity-heat-electricity, heat-electricity or simply storing and using heat without conversion (Denholm et al. 2010). Converting heat (back to) electricity usually reduces efficiency significantly and therefore most applications aim at direct use of heating/cooling capacities. A common application for load shifting and thus VRE integration is using building thermal mass for storing heat or cooling capacities, i.e., by pre-cooling or heating a building during high electricity supply periods (or low price periods) (Reynders et al. 2013; Braun 2003). This approach might be extended (amount of energy and control options) by adding additional capacities, e.g., in underground or rooftops storage tanks (frequently water) (Rankin and Rousseau 2008; Arteconi et al. 2012; Ashok and Banerjee 2003). Within industry, process cooling and heating can be used for applying thermal energy storage with direct use. Examples are also chilled water and ice storage as well as supercritical steam and thermo oil (Oertel 2008).

Flywheels use rotating masses in (close-to) vacuum containments for converting electricity to kinetic energy and back to electricity when needed. Once set up, they exhibit a very long lifetime, low maintenance times, high cycle efficiency (>90 %), little use of hazardous materials and (very) fast response times. However, discharge losses are within several percent per hour and initial investment is significant. Therefore, flywheels are rather used for reserve market applications (Deutsche Energie-Agentur GmbH (dena) 2010; Denholm et al. 2010; Oberhofer 2012). A large-scale flywheel energy storage facility is operated in Stephentown, New York since 2011 (Beacon Power 2015). With 20 MW power and 5 MWh capacity, its main purpose is providing backup capacity.

Supercapacitors (sometimes called ultracapacitors) store energy in an electric field. The absence of moving parts and (direct) energy storage result, among others, in a very high cycle stability (beyond 100,000 cycles), low maintenance cost and relatively high round-trip efficiencies of 80–95 %. However, costs of approx. 320 EUR per kW and a low volumetric energy density of 20 kWh/m^3 prevent large-scale energy storage deployment (Denholm et al. 2010; Deutsche Energie-Agentur GmbH (dena) 2010). Their extremely fast response times make them especially usable for frequency control (Denholm et al. 2010).

Superconducting Magnetic Energy Storage (SMES) uses an electromagnetic field for energy storage. In order to achieve superconductivity, very low temperatures have to be reached and held, which significantly reduced total efficiency. In addition, high investment cost and single-percent discharge losses per day, combined with very fast response times, efficiencies greater than 90 % and low total energy storage determine their main application area within frequency control (Denholm et al. 2010; Oberhofer 2012).

Above energy storage technologies might also be used in combination (hybrid storage technologies), i.e., to achieve fast response times (e.g., capacitor) combined with higher energy density (e.g., from batteries). This approach is, aside from VRE integration, also used in EVs to accommodate short-term high power requests (Neugebauer 2014; Butterbach et al. 2011).

Main challenges of discussed decentralized energy storage options are economically driven: in the presence of grid access and other energy storage options, e.g., pumped hydro, decentralized energy storage needs to be individually evaluated. As such, options which include direct electricity storage or conversion back to electricity have their main application in power quality and backup services applications, but might become favorable options for energy storage in the future. The next section discusses flexibilizing demand to better match supply from a utility perspective, which is another central option to integrate VRE.

2.3.2 *Demand Side Management for Renewable Energy Integration*

In the context of matching electricity demand and supply by influencing the demand side, DSM is a method driven by the supply side (utilities) to influence demand of customers (compare to Fig. 2.16). As such, DSM is closely related to energy flexibility of manufacturing systems (cf. Sect. 2.1.4), but focuses on influencing demand in general (and therefore also includes, for example, energy efficiency and demand shifting). Further, energy flexibility of manufacturing systems is mainly concerned with changing energy demand under given constraints (e.g., technical, operational), while DSM focuses on setting incentives for customers and providing the required operational background to enable customers and utilities to effectively influence demand.

Utilities recognized already more than a hundred years ago that electricity prices should be varied within a day to improve supply economics (Cappers et al. 2010). As such, DSM is a much better covered field than energy flexibility and provides useful insight on methods and tools to influence (electricity) demand. In general, DSM aims at influencing the demand for a good or service by the supply side. DSM is not limited to electricity markets, and, for example, also applied for other infrastructure components, e.g., water (Guy 1996). In the context of electricity markets, DSM can

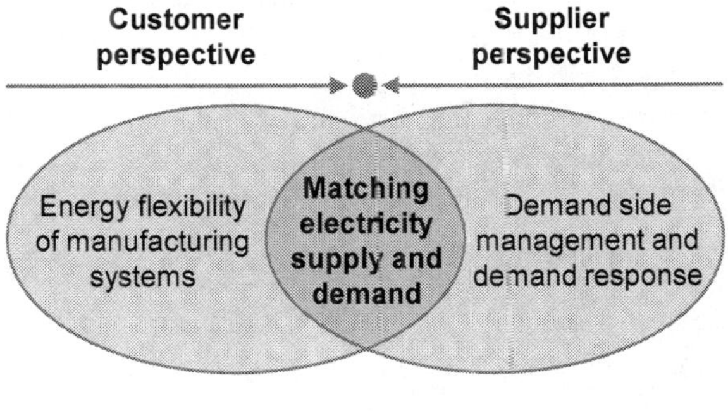

Fig. 2.16 Energy flexibility and demand side management as two perspectives to match electricity supply and demand (own illustration)

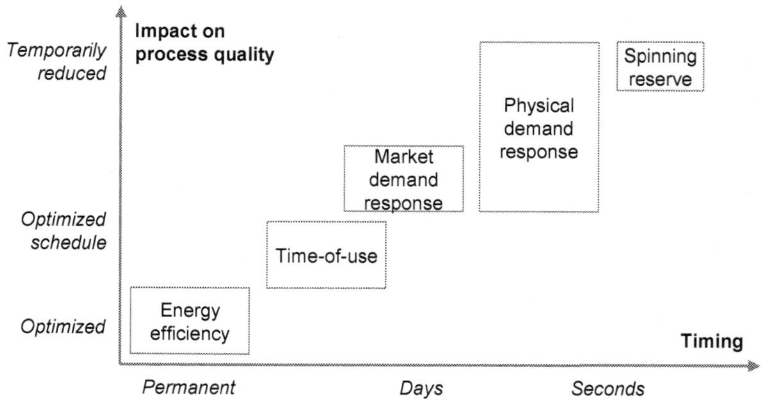

Fig. 2.17 Different categories of demand side management (from Palensky and Dietrich 2011, p. 381, own illustration, original copyright 2011 IEEE)

be defined as "organized utility activities that are intended to affect the amount and timing of customer electricity use" (Electric Power Research Institute 1993, pp. 2–2). Historically, electricity demand volatility was regarded as a given fact by electricity suppliers and distributors, which had to be accommodated by flexible supply. DSM programs are designed to increase demand flexibility, i.e., elasticity of demand, to increase efficiency of electricity systems (electricity suppliers, customers and connecting grid) (Spees and Lave 2007; Aalami et al. 2010). As a general principle to match demand and supply, DSM is a central part of integrating VRE into a power system.

From a timing perspective (duration of impact on load, compare to Fig. 2.17), DSM can be classified into energy efficiency, time of use, (market and physical) demand response, and spinning reserve (Palensky and Dietrich 2011). Depending on the time horizon for upfront notification of a customer by the utility, changes have a different (unwanted) impact on customer processes. From a total energy demand perspective, DSM is categorized into energy efficiency and demand response (Paulus and Borggrefe 2011). Energy efficiency aims at reducing total energy demand, while demand response shifts the timing of demand, leaving overall demand potentially unaffected.

The next two subsections briefly discuss general DSM objectives and then DR as part of DSM. As outlined before, DSM and DR in particular are closely related to energy flexibility of manufacturing systems as they summarize the supply-side perspective on influencing demand. Therefore, the following brief overview highlights relevant objectives and DR methods which can also be found in the context of energy flexibility.

Demand Side Management and Load Shaping Objectives

General load shaping goals are illustrated in Fig. 2.18. As mentioned before, DSM aims at influencing the amount and time of electricity demand. As such, illustrated

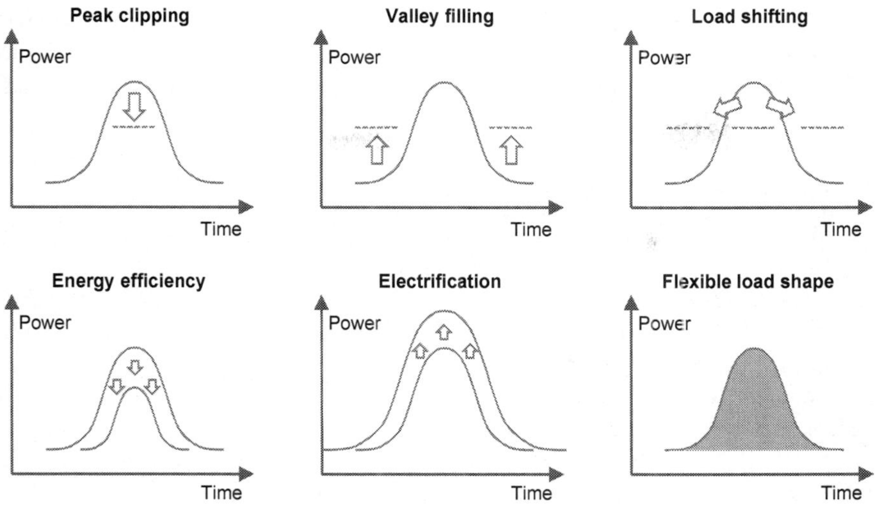

Fig. 2.18 General load shaping goals (own illustration based on Gellings 1985, p. 1469, original copyright 1985 IEEE, Elkarmi and AbuShikhah 2012, p. 102 and Charles River Associates 2005, p. 7)

generic load shaping goals can also be found in reshaping demand of manufacturing systems using energy flexibility, for example, illustrated in Fig. 2.7. A load shape with regards to time can refer to daily, weekly or seasonal changes in demand (Gellings 1985). Goals and corresponding objectives are:

- **Peak clipping** aims at reducing system peak demand to reduce the need for backup generation capacity or other forms of reserve capacity (i.e., direct load control).
- **Valley filling** aims at increasing off-peak electricity demand. This can be beneficial to the utility and customer if long-run costs are lower than electricity price (improved utilization of base-load generation).
- **Load shifting** also aims at reducing peak demand (as peak clipping), but without (substantially) reducing total demand. This is commonly achieved by rescheduling electricity demand, i.e., by using thermal storage capacities (e.g., water heaters) or by using general timing flexibility (e.g., running a sintering machine or electric arc furnace at night).
- **Energy efficiency** aims at (uniformly) decreasing total electricity demand, most common to preserve (natural) resources. Measures include, for example, installing more efficient equipment.
- **Electrification** or load growth increases total electricity demand, for example, to increase utility sales and/or improve load factors of generation plants. Examples include fuel switching (i.e., electric vehicles) and general increase in (electric) energy demand through economic growth.

- **Flexible load shape** refers to a policy where utilities can influence customer demand as needed, commonly to accomplish reliability goals. To achieve a flexible control, the utility needs to be able to dynamically influence customer demand, for example, through direct load control.

Demand Response Strategies and Programs

The idea of DR has been known since the early days of the (U.S.) electric power supply industry. In the late nineteenth century, engineers discussed pricing methods which included different charges dependent on the time of the day in the U.S. (Cappers et al. 2010; Hausman and Neufeld 1984; Isser 2015). Discussed pricing structures aimed at economically efficient allocation of short-term/marginal (generation) costs and long-term costs incurred by providing capacity and required assets to meet different levels of (peak) demand. In the context of VRE integration, DR summarizes a variety of supply-side methods to accommodate increased non-dispatchable VRE share. Without DR, more but less frequently used backup generation capacity might be required to integrate VRE (cf. Sect. 2.2.5, Merit Order Effect).

DR can be defined as "Changes in electric usage by end-use customers from their normal consumption patterns in response to changes in the price of electricity over time, or to incentive payments designed to induce lower electricity use at times of high wholesale market prices or when system reliability is jeopardized" (U.S. Department of Energy 2006, p. 6). The definition differentiates between two customer-side mechanisms for influencing demand, (1) response to price changes or (2) response to incentives, which are commonly used to classify DR programs.

Several different types of DR programs exist. As mentioned, programs are separated into price-based programs or incentive-/event-based programs (Goldman et al. 2010; Albadi and El-Saadany 2007, 2008; U.S. Department of Energy 2006; Elkarmi and AbuShikhah 2012; Charles River Associates 2005):

- **Price-based programs**: Rates for electricity demand are varied over time, inducing an incentive for customers to reduce demand during periods with high prices:
  - Time-of-use (TOU): Different, previously (i.e., for one year) fixed prices depending on time of the day (e.g., multi-hour blocks).
  - Real-time pricing (RTP): Continuously varying prices (hourly or quarter-hourly), dependent on wholesale market.
  - Critical peak pricing (CPP): Mixture of TOU and RTP. Very high rates during a multi-hour CPP event while lower TOU rates during non-CPP events.
  - Extreme day pricing (EDP): Similar to CPP, but critical pricing period is a full day (24 h).
  - Extreme day critical peak pricing (ED-CPP): Very high electricity prices for a number of hours during extreme days (similar to CPP), but otherwise flat.

- **Incentive-/event-based programs**: Customers receive a payment when altering (reducing) loads upon request from the utility:
  - Direct load control: Utilities are enabled to (externally) reduce demand of customers, usually with short upfront notice and only for certain equipment (e.g., water boilers, air conditioners) in exchange for a payment.
  - Demand bidding/buyback programs: Customers bid for reducing loads and receive a monetary compensation if bids are successful (Saebi et al. 2010).
  - Emergency DR: Customers receive a (fixed) compensation during times when load reductions are required and loads are curtailed.
  - Interruptible/curtailable programs: Customers reduce demand upon request from the utility in return for lower rates.
  - Capacity market programs: Load reductions are offered by customers to substitute for system capacity (i.e., conventional generation or distribution capacity). Customers commit (typically day-ahead) to reduce loads during a prespecified time period and receive a payment in return, also when load reductions are not called. In return, customers face penalties if requested reductions are not achieved.
  - Ancillary services market programs: Customers offer to cut loads on a market typically operated by an ISO/regional transmission organization (RTO) and therefore act as operating reserves. If bids are successful, payments are received for providing the option to reduce loads upon request from the operator.

Empirical evidence for DR impact and potential has been surveyed, for example, in Cappers et al. (2010). Existing peak load reduction potential through DR programs was estimated to be in a single-digit percentage range in the U.S. in 2008 and increasing. Although the concept of DR is not new, implementation in electricity markets is still slow. In most markets, utilities charge customers an energy and a demand rate dependent on their electricity usage pattern rather than based on their demand in relation to current system load.

2.4 Intermediate Summary and Conclusion

This chapter starts with relevant background on manufacturing systems. In order to provide the required detail on the energy/electricity demand side, energy demand of production is discussed, including energy flexibility as a novel approach towards VRE integration from a demand side perspective. The second section highlights the energy supply side and includes background on conventional and renewable energy generation (compare also to Fig. 2.9 for an overview of highlighted topics). Characteristics and emerging challenges from renewable and, as most renewable generation is locally spread out, specifically decentralized generation are discussed. The connecting elements of electricity demand and supply, electricity distribution, markets and related grid stability challenges due to VRE integration, are then briefly summarized. The third section details specific options for integrating VRE into power systems, starting with energy storage options. Subsequently, DSM and DR in particular are discussed, as both areas summarize approaches for VRE integration from a supplier (utility) perspective. Due to their relatively long history, these approaches can provide valuable insight into methods for demand and supply matching. Further, as outlined before, DSM and DR methods are closely related to energy flexibility, which is the equivalent, thus relatively young (compared to DSM) area from a demand side (industry) perspective. The objective stated in Chap. 1 aims at developing a concept for VRE integration from a demand side and thus energy flexibility perspective. As such, existing research in the field of energy flexible manufacturing systems is analyzed in the next chapter.

References

Aalami, H., Moghaddam, M. P., & Yousefi, G. (2010). Modeling and prioritizing demand response programs in power markets. *Electric Power Systems Research, 80*, 426–435.

AG Energiebilanzen e.V. (2015). Bruttostromerzeugung in Deutschland ab 1990 nach Energieträgern. Technical report, Berlin. http://www.ag-energiebilanzen.de/index.php?article_id=29&fileName=20151112_brd_stromerzeugung1990-2014.pdf.

Albadi, M. H., & El-Saadany, E. F. (2007). Demand response in electricity markets: An overview. *IEEE Power & Energy Society General Meeting PES* (pp. 1–5).

Albadi, M. H., & El-Saadany, E. F. (2008). A summary of demand response in electricity markets. *Electric Power Systems Research, 78*, 1989–1996.

Allwood, J. M., & Cullen, J. M. (2012). *Sustainable materials: With both eyes open*. Cambridge: UIT.

Allwood, J. M., Ashby, M. F., Gutowski, T. G., & Worrell, E. (2011). Material efficiency: A white paper. *Resources, Conservation and Recycling, 55*(3), 362–381.

Apostolos, F., Alexios, P., Georgios, P., Panagiotis, S., & George, C. (2013). Energy efficiency of manufacturing processes: A critical review. *Procedia CIRP, 7*, 628–633. http://linkinghub.elsevier.com/retrieve/pii/S2212827113003132.

Arteconi, A., Hewitt, N., & Polonara, F. (2012). State of the art of thermal storage for demand-side management. *Applied Energy, 93*, 371–389.

Ashok, S., & Banerjee, R. (2003). Optimal cool storage capacity for load management. *Energy, 28*(2), 115–126.

Beach, R., Muhlemann, A. P., Price, D. H. R., Paterson, A., & Sharp, J. A. (2000). A review of manufacturing flexibility. *European Journal of Operational Research, 122*, 41–57. http://www.sciencedirect.com/science/article/pii/S0377221799000624.

Beacon Power (2015). Beacon power. Retrieved November 13, 2015 from http://beaconpower.com/.

Beier, J., Thiede, S., & Herrmann, C. (2016). Integrating variable renewable electricity supply into manufacturing systems. In C. Herrmann & S. Kara (Eds.), *Eco-factories of the future (accepted manuscript)*. New York: Springer.

Bernal-Agustín, J. L., & Dufo-López, R. (2009). Simulation and optimization of stand-alone hybrid renewable energy systems. *Renewable & Sustainable Energy Reviews, 13*(8), 2111–2118.

Braun, J. E. (2003). Load control using building thermal mass. *Journal of Solar Energy Engineering, 125*(3), 292–301.

Bullough, C., Gatzen, C., Jakiel, C., Koller, M., Nowi, A., & Zunft, S. (2004). Advanced adiabatic compressed air energy storage for the integration of wind energy. In *Proceedings of the European Wind Energy Association Conference EWEC* (pp. 22–25). London.

Bundesministerium für Wirtschaft und Energie (BMWi) (2013). Langfristige Steuerung der Versorgungssicherheit im Stromsektor. Technical report, Bundesministerium für Wirtschaft und Energie, Berlin.

Bundesnetzagentur (2014). Feststellung des Reservekraftwerksbedarfs für den Winter 2014/2015 sowie die Jahre 2015/2016 und 2017/2018 und zugleich Bericht über die Ergebnisse der Prüfung der Systemanalysen. Technical report, Bonn.

Butterbach, S., Vulturescu, B., Forgez, C., Coquery, G., & Friedrich, G. (2011). Lead-acid battery model for hybrid energy storage. In *IEEE Vehicle Power and Propulsion Conference* (pp. 1–5). Chicago, IL: IEEE.

Buzacott, J. A., & Mandelbaum, M. (1985). Flexibility and productivity in manufacturing systems. In *Proceedings of the Annual IIE Conference* (pp. 404–413). Chicago, IL.

Cappers, P., Goldman, C., & Kathan, D. (2010). Demand response in U.S. electricity markets: Empirical evidence. *Energy*, *35*(4), 1526–1535. http://dx.doi.org/10.1016/j.energy.2009.06.029.

Charles River Associates (2005). Primer on demand-side management: With an emphasis on price-responsive programs. Technical report, Oakland, CA.

Chryssolouris, G. (1992). *Manufacturing systems: Theory and practice*. New York, NY: Springer.

Crotogino, F., Mohmeyer, K.-U., & Scharf, R. (2001). Huntorf CAES: More than 20 years of successful operation. In *Solution Mining Research Institute, Spring Meeting* (pp. 351–357). Orlando, FL.

Daryanian, B., Bohn, R. E., & Tabors, R. D. (1989). Optimal demand-side response to electricity spot prices for storage-type customers. *IEEE Transactions on Power Systems*, *4*(3), 897–903. http://ieeexplore.ieee.org/lpdocs/epic03/wrapper.htm?arnumber=4310873.

Denholm, P., Ela, E., Kirby, B., & Milligan, M. (2010). The role of energy storage with renewable electricity generation. Technical report, National Renewable Energy Laboratory, Golden, CO. http://www.nrel.gov/docs/fy10osti/47187.pdf.

DESTATIS (2014). Industriebetriebe produzieren knapp 9% der in Deutschland erzeugten Strommenge. Retrieved March 16, 2015 from https://www.destatis.de/DE/PresseService/Presse/Pressemitteilungen/2014/10/PD14_363_433.html.

DESTATIS (2015). Energieverwendung der Betriebe im Verarbeitenden Gewerbe 2013. Retrieved March 16, 2015 from https://www.destatis.de/DE/ZahlenFakten/Wirtschaftsbereiche/Energie/Verwendung/Tabellen/KohleErdgasStrom.html.

Deutsche Energie-Agentur GmbH (dena) (2010). Dena grid study II. Integration of renewable energy sources in the German power supply system from 2015-2020 with an outlook to 2025. Technical report, Berlin.

DIHK and VEA (2014). Faktenpapier Eigenerzeugung von Strom: Rahmenbedingungen, Trends, Beispiele. Technical report, Berlin, Brüssel, Hannover.

Duflou, J. R., Sutherland, J. W., Dornfeld, D., Herrmann, C., Jeswiet, J., & Kara, S. (2012). Towards energy and resource efficient manufacturing: A processes and systems approach. *CIRP Annals - Manufacturing Technology*, *61*(2), 587–609. http://linkinghub.elsevier.com/retrieve/pii/S0007850612002016.

Dyckhoff, H. (2006). *Produktionstheorie: Grundzüge industrieller Produktionswirtschaft* (5th ed.). Berlin, Heidelberg, New York: Springer.

Dyckhoff, H., & Spengler, T. S. (2010). *Produktionswirtschaft: Eine Einführung* (3rd ed.). Heidelberg, Dordrecht, London, New York: Springer.

EEX (2014a). Transparency in energy markets - Actual solar power generation (table). Retrieved March 29, 2014 from http://www.eex-transparency.com/homepage/power/germany/production/usage/actual-solar-power-generation-/actual-solar-power-generation-table-

EEX (2014b). Transparency in energy markets - Actual wind power generation (table). Retrieved March 29, 2014 from http://www.eex-transparency.com/homepage/power/germany/production/usage/actual-wind-power-generation/actual-wind-power-generation-table-.

EEX (2015). EEX homepage. Retrieved December 04, 2015 from https://www.eex.com/en/.

Electric Power Research Institute (1993). Principles and practice of demand side management. Technical report, Palo Alto, CA.

Elkarmi, F., & AbuShikhah, N. (2012). *Power system planning technologies and applications: Concepts, solutions and management*. Hershey, PA: Engineering Science Reference.
Ellabban, O., Abu-Rub, H., & Blaabjerg, F. (2014). Renewable energy resources: Current status, future prospects and their enabling technology. *Renewable & Sustainable Energy Reviews, 39*, 748–764. http://dx.doi.org/10.1016/j.rser.2014.07.113.
ENTSO-E (2009). P1: Load-frequency control and performance. *Continental Europe Operation Handbook* (pp. P1-1–P1-32).
ENTSO-E (2014). Hourly load values for a specific country for a specific month (in MW). Retrieved March 31, 2014 from https://www.entsoe.eu/db-query/consumption/mhlv-a-specific-country-for-a-specific-month.
ENTSO-E (2015). European network of transmission system operators for electricity. Retrieved December 04, 2015 from https://www.entsoe.eu/.
EPEX SPOT (2014). EPEX SPOT SE: Day-ahead auction. Retrieved March 29, 2014 from http://www.epexspot.com/en/market-data.
Gallagher, K. G., & Nelson, P. A. (2014). Manufacturing costs of batteries for electric vehicles. In G. Pistoia (Ed.), *Lithium-ion batteries: Advances and applications* (pp. 97–126). Amsterdam: Elsevier B.V.
Geitmann, S. (2010). *Erneuerbare Energien: Mit neuer Energie in die Zukunft*. Oberkrämer: Hydrogeit.
Gellings, C. W. (1985). The concept of demand-side management for electric utilities. *Proceedings of the IEEE, 73*(10), 1468–1470.
Germany Trade and Invest (2015). The energy storage market in Germany. Technical report, Köln.
Gildemeister energy solutions GmbH (2016). Intelligente Speichersysteme auf Vanadium-Redox-Flow Technologie. Retrieved March 21, 2016 from http://energy.gildemeister.com/de/speichern/cellcube-fb-10-20-30#uebersicht.
Goldman, C., Reid, M., Levy, R., & Silverstein, A. (2010). Coordination of energy efficiency and demand response. Technical report, Lawrence Berkeley National Laboratory. http://escholarship.org/uc/item/1z43c7bt.pdf.
Graßl, M., Vikdahl, E., & Reinhart, G. (2013). A petri-net based approach for evaluating energy flexibility of production machines. In M. Zäh (Ed.), *5th International Conference on Changeable, Agile, Reconfigurable and Virtual Production (CARV 2013)* (pp. 303–308). Switzerland, Munich: Springer International Publishing.
Günther, H.-O., & Tempelmeier, H. (2005). *Produktion und Logistik* (6th ed.). Berlin, Heidelberg, New York: Springer.
Gupta, Y. P., & Goyal, S. (1989). Flexibility of manufacturing systems: Concepts and measurements. *European Journal of Operational Research, 43*, 119–135. http://www.sciencedirect.com/science/article/pii/0377221789902063.
Gutowski, T., Murphy, C., Allen, D., Bauer, D., Bras, B., & Piwonka, T. (2005). Environmentally benign manufacturing: Observations from Japan, Europe and the United States. *Journal of Cleaner Production, 13*(1), 1–17. http://linkinghub.elsevier.com/retrieve/pii/S0959652603001781.
Gutowski, T. G., Dahmus, J., & Thiriez, A. (2006). Electrical energy requirements for manufacturing processes. In *13th CIRP International Conference of Life Cycle Engineering* (pp. 623–627). Lueven.
Guy, S. (1996). Managing water stress: The logic of demand side infrastructure planning. *Journal of Environmental Planning and Management, 39*(1), 123–130.
Hausman, W. J., & Neufeld, J. L. (1984). Time-of-day pricing in the U.S. electric power industry at the turn of the century. *The RAND Journal of Economics, 15*(1), 116–126.
Herrmann, C., & Thiede, S. (2009). Process chain simulation to foster energy efficiency in manufacturing. *CIRP Journal of Manufacturing Science and Technology, 1*(4), 221–229. http://linkinghub.elsevier.com/retrieve/pii/S1755581709000121.
Herrmann, C., Schmidt, C., Kurle, D., Blume, S., & Thiede, S. (2014). Sustainability in manufacturing and factories of the future. *International Journal of Precision Engineering and Manufacturing-Green Technology, 1*(4), 283–292. http://link.springer.com/10.1007/s40684-014-0034-z.

Hesselbach, J., Herrmann, C., Detzer, R., Martin, L., Thiede, S., & Lüdemann, B. (2008). Energy efficiency through optimized coordination of production and technical building services. In *LCE2008 - 15th CIRP International Conference on Life Cycle Engineering* (pp. 624–628). Sydney.

Icha, P. (2015). Entwicklung der spezifischen Kohlendioxid-Emissionen des deutschen Strommix in den Jahren 1990 bis 2014. Technical report, Umweltbundesamt, Dessau-Roßlau.

IPCC (2011). IPCC Special report on renewable energy sources and climate change mitigation. Prepared by working group III of the intergovernmental panel on climate change. In O. Edenhofer, R. Pichs-Madruga, Y. Sokona, K. Seyboth, P. Matschoss, S. Kadner, T. Zwickel, P. Eickemeier, G. Hansen, S. Schlömer & C. von Stechow (Eds.), Cambridge, United Kingdom and New York, NY, USA: Cambridge University Press.

IPCC (2014). Climate change 2014: Mitigation of climate change. Working group III contribution to the IPCC 5th assessment report: Chapter 7: Energy systems. In *Climate Change 2014: Mitigation of Climate Change. Working Group III Contribution to the IPCC 5th Assessment Report.* Cambridge, New York, NY: Cambridge University Press.

Isser, S. (2015). *Electricity restructuring in the United States: Markets and policy from the 1978 energy act to present.* New York, NY: Cambridge University Press.

Kabelitz, S., & Streckfuß, U. (2014). Energieflexibilität in der Produktionstheorie. *ZWF Zeitschrift für wirtschaftlichen Fabrikbetr, 109*(1–2), 43–45.

Kara, S., & Li, W. (2011). Unit process energy consumption models for material removal processes. *CIRP Annals - Manufacturing Technology, 60*(1), 37–40. http://linkinghub.elsevier.com/retrieve/pii/S0007850611000199.

Karellas, S., & Tzouganatos, N. (2014). Comparison of the performance of compressed-air and hydrogen energy storage systems: Karpathos island case study. *Renewable & Sustainable Energy Reviews, 29*, 865–882.

Kosmadakis, G., Karellas, S., & Kakaras, E. (2013). Renewable and conventional electricity generation systems: Technologies and diversity of energy systems. In J. M. Hills & E. Michalena (Eds.), *Renewable Energy Governance* (pp. 9–30). London: Springer.

Liebl, C., Popp, R., & Zäh, M. F. (2015). Energieflexibilität von Produktionsanlagen. *wt Werkstattstech online, 105*(3), 136–140.

Lorenz, S., Putz, M., & Schlegel, A. (2012). Energieeffizienz 2.0: Neue Geschäftsmodelle auch für die Industrie. *ZWF Zeitschrift für wirtschaftlichen Fabrikbetr, 107*(9), 599–602.

Lund, H., & Salgi, G. (2009). The role of compressed air energy storage (CAES) in future sustainable energy systems. *Energy Conversion and Management, 50*(5), 1172–1179.

Lund, H., Salgi, G., Elmegaard, B., & Andersen, A. N. (2009). Optimal operation strategies of compressed air energy storage (CAES) on electricity spot markets with fluctuating prices. *Applied Thermal Engineering, 29*(5–6), 799–806.

Mascarenhas, B. (1981). Planning for flexibility. *Long Range Planning, 14*(5), 78–82. http://www.sciencedirect.com/science/article/pii/002463018190011X.

Measurlogic (2015). Maximum demand control systems. Technical report. http://www.measurlogic.com/EnergyManagement/Resources/MaxDemandBook_loM6.pdf.

Neugebauer, R. (2014). *Handbuch Ressourcenorientierte Produktion.* KG, Wien: Carl Hanser Verlag GmbH und Co.

Nielsen, L., & Leithner, R. (2009). Dynamic simulation of an innovative compressed air energy storage plant - Detailed modelling of the storage cavern. *WSEAS Transactions on Power Systems, 4*(8), 253–263.

Nyhuis, P., Fronia, P., Pachow-Frauenhofer, J., & Wulf, S. (2009). Wandlungsfähige Produktionssysteme. *wt Werkstattstech online, 99*(4), 205–210.

Oberhofer, A. (2012). Energy storage technologies and their role in renewable integration. Technical report, Global Energy Network Institute. https://www.geni.org/globalenergy/research/energy-storage-technologies/Energy-Storage-Technologies.pdf.

Oertel, D. (2008). Energiespeicher - Stand und Perspektiven: Sachstandsbericht zum Monitoring "Nachhaltige Energieversorgung". Technical report, TAB - Büro für Technikfolgen-Abschätzung beim Deutschen Bundestag, Berlin.

Palensky, P., & Dietrich, D. (2011). Demand side management: Demand response, intelligent energy systems, and smart loads. *IEEE Transactions on Industrial Informatics*, *7*(3), 381–388.
Paulus, M., & Borggrefe, F. (2011). The potential of demand-side management in energy-intensive industries for electricity markets in Germany. *Applied Energy*, *88*(2), 432–441.
Perman, R., Ma, Y., McGilvray, J., & Common, M. (2003). *Natural resource and environmental economics* (3rd ed.). Harlow (Essex).
Popp, R., & Zäh, M. F. (2014). Steuerung des Energiebedarfs von Werkzeugmaschinen. *wt Werkstattstech online*, *104*(6), 413–417.
Popp, R., Keller, F., Atabay, D., Dornmair, R., Buderus, J., & Kohl, J. (2013). Technische Innovationen für die Energieflexible Fabrik. *ZWF Zeitschrift für wirtschaftlichen Fabrikbetr*, *108*(07–08), 556–560.
Posselt, G. (2016). *Towards energy transparent factories*. Cham: Springer International Publishing.
Qazi, A., Fayaz, H., Wadi, A., Raj, R. G., & Rahim, N. (2015). The artificial neural network for solar radiation prediction and designing solar systems: A systematic literature review. *Journal of Cleaner Production*, *104*, 1–12.
Quaschning, V. (2011). *Regenerative Energiesysteme: Technologie - Berechnung - Simulation* (8th ed.). München: Hanser Verlag.
Rankin, R., & Rousseau, P. (2008). Demand side management in South Africa at industrial residence water heating systems using in line water heating methodology. *Energy Conversion and Management*, *49*(1), 62–74.
Rehman, S., Al-Hadhrami, L. M., & Alam, M. M. (2015). Pumped hydro energy storage system: A technological review. *Renewable & Sustainable Energy Reviews*, 44:586–598. http://linkinghub.elsevier.com/retrieve/pii/S1364032115000106.
Reinhart, G., & Schultz, C. (2014). Herausforderungen einer energieorientierten Produktionssteuerung. *ZWF Zeitschrift für wirtschaftlichen Fabrikbetr*, *109*(1–2), 29–33.
Reynders, G., Nuytten, T., & Saelens, D. (2013). Potential of structural thermal mass for demand-side management in dwellings. *Building and Environment*, *64*, 187–199.
Saebi, J., Taheri, H., Mohammadi, J., & Nayer, S. S. (2010). Demand bidding/buyback modeling and its impact on market clearing price. In *IEEE International Energy Conference and Exhibition (ENERGYCON)* (pp. 791–796).
Schuh, G. (2006). *Produktionsplanung und -steuerung: Grundlagen, Gestaltung und Konzepte* (3rd ed.). Berlin, Heidelberg, New York: Springer.
Schuh, G., & Stich, V. (2012). *Produktionsplanung und -steuerung 1: Grundlagen der PPS* (4th ed.). Berlin/Heidelberg: Springer Vieweg.
Schwab, A. J. (2015). *Elektroenergiesysteme* (4th ed.). Heidelberg: Springer Vieweg.
Sensfuß, F., Ragwitz, M., & Genoese, M. (2008). The merit-order effect: A detailed analysis of the price effect of renewable electricity generation on spot market prices in Germany. *Energy Policy*, *36*(8), 3076–3084.
Slack, N. (1983). Flexibility as a manufacturing objective. *International Journal of Operations & Production Management*, *3*(3), 4–13.
Spees, K., & Lave, L. B. (2007). Demand response and electricity market efficiency. *The Electricity Journal*, 20(3):69–85. http://www.sciencedirect.com/science/article/pii/S1040619007000188.
Thiede, S. (2012). *Energy efficiency in manufacturing systems*. Berlin, Heidelberg: Springer.
U.S. Department of Energy (2006). Benefits of demand response in electricity markets and recommendations for achieving them: A report to the United States congress pursant to section 1252 of the energy policy act of 2005. Technical report, U.S. Department of Energy. http://energy.gov/sites/prod/files/oeprod/DocumentsandMedia/DOE_Benefits_of_Demand_Response_in_Electricity_Markets_and_Recommendations_for_Achieving_Them_Report_to_Congress.pdf.
Weinert, N. (2010). *Vorgehensweise für Planung und Betrieb energieeffizienter Produktionssysteme*. Stuttgart: Fraunhofer Verlag.
Weinert, N., Chiotellis, S., & Seliger, G. (2011). Methodology for planning and operating energy-efficient production systems. *CIRP Annals - Manufacturing Technology*, *60*(1), 41–44. http://linkinghub.elsevier.com/retrieve/pii/S0007850611000163.

Westkämper, E. (2006). *Einführung in die Organisation der Produktion*. Berlin, Heidelberg, New York: Springer.

Wiendahl, H.-P. (2014). *Betriebsorganisation für Ingenieure* (8th ed.). München: Hanser Verlag.

Zäpfel, G. (2001). *Grundzüge des Produktions- und Logistikmanagements* (2nd ed.). München, Wien: Oldenbourg.

Zunft, S., Jakiel, C., Koller, M., & Bullough, C. (2006). Adiabatic compressed air energy storage for the grid integration of wind power. In *Sixth International Workshop on Large-Scale Integration of Wind Power and Transmission Networks for Offshore Windfarms* (pp. 1–6). Delft.

Chapter 3
Existing Approaches in the Field of Energy Flexible Manufacturing Systems

The formulated bottom-up research question from Chap. 1 is focused on methods and tools towards flexible electricity demand of industry. The second chapter described theoretical background on manufacturing systems, their energy demand and energy flexibility. Consequently, the next step is to review, classify, and evaluate existing approaches to identify what methods towards fulfillment of mentioned research question are already available and to analyze potential gaps. As outlined in the previous chapter, described research question is closely related to the topic of energy flexibility, which is selected as the focal point for identifying existing methods. First, general considerations for selecting and classifying approaches are described, which also covers criteria for including and excluding approaches for detailed review. Selected publications are then reviewed in detail. Detailed evaluation criteria are derived from presented theoretical background (Chap. 2) and analyzed publications. This evaluation framework is then applied to previously summarized approaches. The result is a comprehensive overview of the current state of existing research on energy flexibility, with a focus on manufacturing. Based on this, gaps are identified and prevailing research demand is derived.

3.1 Selection and Classification Method

In order to identify relevant existing approaches, a detailed review of existing research has been conducted. Publications from manufacturing and energy engineering as well as operations research (OR) have been the main search areas. As the approach presented in this work is built on several disciplines, relevant related and/or referenced publications and technical reports were also considered.

J. Beier, *Simulation Approach Towards Energy Flexible Manufacturing Systems*, Sustainable Production, Life Cycle Engineering and Management, DOI 10.1007/978-3-319-46639-2_3

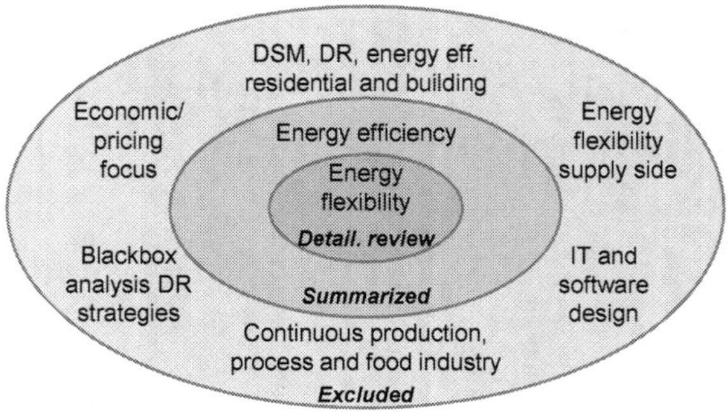

Fig. 3.1 Research selection logic (own illustration)

Due to the fact that several disciplines are related to energy flexibility in manufacturing (i.e., energy and production engineering, control theory), the number of approaches which are reviewed in detail needs to be reduced to focus on the most relevant ones in detail. Figure 3.1 summarizes the logic which was applied to select relevant research. Although related to energy flexibility, several approaches are excluded from a detailed review as they are relatively distant from the specific topic in focus (outer circle). Energy efficiency centered approaches, which are relevant for developing a novel energy flexibility concept, are briefly summarized (middle circle). Approaches which fall within the inner circle, directly energy flexibility related approaches, are reviewed in detail.

Starting with the outer circle and with regards to the objective outlined in Chap. 1 and classification based on current state of the art (Chap. 2), the following criteria were used to exclude approaches from detailed discussion:

- **Energy flexibility on the supply side**: Approaches which are centered around flexible scheduling of (decentralized) electricity generation equipment (i.e., to reduce peak demand), without specific focus on the electricity demand side, are not included for detailed review. Examples include Ashok and Banerjee (2003b), Babu and Ashok (2009), Mitra et al. (2013) (with an extensive overview of combined heat an power (CHP) plant scheduling) and Ghadimi et al. (2015). An example for converting DR into a virtual generation capacity can be found in Kwag and Kim (2012).
- **DSM, DR, and energy efficiency for residential and building sector**: Approaches which aim at applying DSM techniques and objectives to building and household equipment (i.e., water boilers, washing machines, refrigerators, and air conditioning (AC)) are not considered as part of the relevant focus area. Examples include Reynders et al. (2013), Rankin and Rousseau (2008), Finn et al. (2013), Ha et al. (2012) and Braun (2003), for an overview and state of the art of DSM utilizing thermal storage see, e.g., Arteconi et al. (2012).
- **Continuous production, process, and food industry**: Methods and approaches which focus on continuous production plants (i.e., production systems not involving or which cannot be directly related to a manufacturing environment), for example found within the process and food industry, are not reviewed in detail. This

includes production systems primarily found in process industry involving mainly continuous and/or batch-type processes. Energy flexibility and DR-related examples include Castro et al. (2009, 2011), Mitra et al. (2012), and van Staden et al. (2011). A food production-related example (utilizing cooling capacities during meat production) can be found in Alcázar-Ortega et al. (2012a, b). An early example for production scheduling under TOU rates using nonlinear, generic electricity demand functions (similar to process industry/continuous process scheduling) is described in Nilsson (1992) and Nilsson and Söderström (1993). As manufacturing and continuous process plants exhibit substantially different characteristics, the scope of this work is limited to manufacturing plants to reduce complexity. Although this criterion narrows the research review, the proposed concept will be evaluated against its applicability for mixed discrete/continuous process chains.
- **Economic/pricing focus**: Theory and approaches which are mainly concerned with optimal electricity pricing dependent on electricity supply and demand (including changes in demand dependent on prices) are excluded (e.g., Outhred et al. 1988).
- **Black box analysis of DR strategies**: Studies which provide no relevant insight on how DR strategies, actions, and policies are actually implemented are not included in a detailed summary. For example, a descriptive study on effects of varying electricity prices, related industry response and resulting increase of wind energy demand in Ireland (as a consequence of the Merit Order Effect) can be found in Finn and Fitzpatrick (2014). Statistical evidence on effectiveness of DR is provided, but no further detail on how the two studied facilities (a cold storage facility and a manufacturing plant) achieved load shifting. Another example can be found in Jang et al. (2015), where a comprehensive analysis of industrial and commercial industries from Korea and their responses to a CPP program is performed to guide design and implementation of DR programs.
- **IT and software design**: Methods which focus on software design (requirements, implementation, database structures, etc.) are excluded from detailed review unless they also include additional energy flexibility related methodological content. Examples include Bougain et al. (2015) and Gerhard (2015).

Although above criteria are used to exclude identified approaches from detailed review to limit the total number of reviewed publications, their content and findings are still considered for (a) developing a detailed evaluation scheme for existing research and (b) for developing the concept outlined in the following. In particular, selected publications which have been excluded for detailed review are referenced and briefly described if they contain relevant information, for example, if an additional application example or implementation method is provided.

The next two sections highlight existing DSM-related research. As such, the first section outlines energy efficiency, the second section energy flexibility centered approaches.

3.2 Energy Efficiency Centered Approaches

Although energy efficiency is closely related to energy flexibility and a central part of DSM, a comprehensive review of energy efficiency including background, current research topics, and subsequently resulting research demand would exceed the scope in relation to the focus topic. Instead, selected existing publications are briefly summarized to provide a basic overview of currently discussed energy efficiency topics. Subsequently, energy efficiency approaches which use load-shifting-related methods or mention their potential extension towards energy flexibility are described in more detail.

Identifying influencing factors and predicting energy demand of processes has been a focus area of research towards energy-efficient manufacturing processes.[1] For example, Gutowski et al. (2006) use statistical analysis to describe specific energy demand and find an inverse relation to the process' material removal rate (MRR). Other examples for analyzing, forecasting, and thus identifying potential for energy demand reduction of manufacturing processes (e.g., milling, turning, and grinding) can be found in Kara and Li (2011), Li and Kara (2011), Balogun and Mativenga (2013), Diaz et al. (2012), Li et al. (2013), Velchev et al. (2014) and Liu et al. (2015). In order to increase energy and, in general, ecological efficiency of processes, eco-efficiency was included in Li et al. (2012c). A detailed methodology to move towards reduced energy demand of machine tools can be found in Zein (2012). On a factory level, integrated monitoring of energy and material flows has been acknowledged as main prerequisite for energy efficiency improvement (Ghadimi et al. 2014; Posselt 2016). Energy demand monitoring and simulation has been extended by including optimization methods and machine control to improve and implement energy efficiency of production machines (Abele et al. 2015a, b; Schlechtendahl et al. 2016; Eberspächer et al. 2014). An approach to evaluate environmental impacts on different hierarchical levels of manufacturing systems is presented in Heinemann et al. (2014). Note that numerous other publications can be found which are centered around energy efficiency on different system levels and which apply several other methods. However, as energy efficiency is not the focus topic, no additional publications are mentioned here.

Several energy efficiency centered approaches can be found which mention or are related to flexible energy demand and energy-related production scheduling and control. The following overview includes approaches which also mention a potential applicability towards load shifting, or apply load shifting related methods to achieve their individual goals (sorted by year and then alphabetically). A critical review is only provided if necessary for highlighting why a given approach is not included in the detailed review (and thus within the central circle in Fig. 3.1). As already several energy flexibility centered approaches have been identified which are closer to the focus topic than the following energy efficiency approaches, a critical review

[1] Note that the structure and references of this paragraph are based on Beier et al. (2017).

of each following energy efficiency approach would exceed the scope of this work. However, they are summarized to give a comprehensive overview of all relevant energy flexibility related, thus not energy flexibility centered approaches, as they provide valuable insight for developing a new energy flexibility concept[2]:

- A hybrid flow shop scheduling problem to minimize energy (electricity) demand is suggested in Liu et al. (2008), proposing a mixed integer nonlinear programming (MINLP) model. Energy demand is minimized, while makespan is modeled as a constraint to fulfill production targets. A genetic algorithm is modified and used to obtain a solution.
- An optimization method is proposed in Rager (2008) to schedule identical, parallel, single-stage machines to reduce energy demand. A detailed quantitative decision support model based on a machine scheduling problem, which also includes energy costs, is proposed. A focus is set on minimizing energy demand and thus costs. Related to peak power demand, a leveling of energy demand is pursued. The approach is applied to an example from the textile industry. Overall, the presented approach largely focuses on the mathematical solution of formulated decision problem.
- On the system and planning level, a method for estimating energy demand of production systems during planning is described in Weinert (2010) and Weinert et al. (2011). The method uses decomposition of total energy demand into different (individual) energy demand segments (*EnergyBlocks*) with fixed and variable duration. The method includes a reference to variable (daily fluctuating) energy prices and to the method's applicability towards reflection of price variability in production planning.
- Another approach includes peak energy demand into the flow shop scheduling problem as objective (Fang et al. 2011). Different processing speeds are considered. Nonetheless, the resulting mixed integer linear programming (MILP) formulation with multiple objectives has a computational complexity and time which does not allow application at an industrial scale.
- The authors in Pechmann and Schöler (2011) propose a method to define energy-related key performance indicators (KPIs) for production systems with a special focus on production planning and control (called *E-PPS*). An eight-step method is proposed which enables the user to develop a set of KPIs and thus evaluate effectiveness of measures which aim at reducing energy and peak electricity demand. Integrating these KPIs into a production control and shifting start dates of processing equipment is claimed to reduce energy and peak demand. The focus is set towards the theoretical description of the KPI development process, neither specific KPIs nor a concrete method for load shifting is provided.
- In Bruzzone et al. (2012), a mixed integer programming (MIP) problem is formulated to reduce peak electricity demand in a flexible flow shop environment. In order to reduce peak demand, production schedules are optimized towards reducing peaks while maintaining operational goals such as job sequencing. However,

[2]Partly based on Beier et al. (2017).

the authors note that some operational objectives, e.g., tardiness and makespan, might worsen. A heuristic solution method is used due to computational complexity of the problem.
- Considering production lines and utilizing simulation, a method has been proposed which determines machines which can be switched to a state with lower power demand due to blocking or starvation in Li et al. (2012b).
- A manufacturing execution-related approach for energy-aware production scheduling and control is presented in Neugebauer et al. (2012) and Langer et al. (2014). Several different tools for energy-sensitive MES have been developed, among them visualization environments, production control systems, data storage options, and a simulation model to demonstrate application of proposed concepts. The central focus is set towards energy efficiency improvement. Within the context of energy-sensitive order scheduling, the authors suggest to use scheduling flexibility of manufacturing and related (e.g., HVAC) processes to take advantage of varying electricity prices or to accommodate limited available electricity without providing a specific method or solution (Neugebauer et al. 2012). Energy efficiency related work includes strategies which utilize intermediate buffer inventories to switch production machines off or control of operating states to achieve energy savings. An IT-concept and software prototype including simulation for verification purposes is described in Langer et al. (2014).
- A holistic approach which considers state-based energy demand of production processes, their dynamic interaction with other processes and auxiliary system(s) to predict dynamic energy demand of production systems through simulation is detailed in Thiede (2012). The proposed energy demand model is used to study the impact of different energy efficiency measures and to reveal potential dynamic/second-order effects.
- A simulation-based approach for energy efficiency improvement of manufacturing systems utilizing production scheduling flexibility is presented in Junge (2007). Production control, in particular order sequencing, is used to reduce energy demand of a production system. A focus is set on ventilation and related heat losses. Three optimization strategies are presented to reduce energy requirements, e.g., through emission control and thus reduced heat losses due to ventilation: (a) scheduling according to a temperature forecast, (b) scheduling dependent upon day or night and (c) scheduling dependent upon ventilation heat losses. The method is also briefly applied in a context of hourly fluctuating electricity prices and limited, time-dependent electricity availability by adapting the ventilation heat loss scheduling method.
- In Dai et al. (2013), a quantitative approach for flexible flow shop scheduling towards improved energy efficiency is proposed. Several technical parameters are considered. An increase in efficiency is, e.g., achieved through avoiding excessive idle electricity demand. The authors propose a modified genetic simulation annealing algorithm to solve their quantitative model.
- An approach developed by Li and Sun (2013) uses Markov Decision Process (MDP) to determine which production machines in a serial production line can be switched off using inventory in buffers to compensate for production stops

of machines. The acquired solution is near-optimal, throughput reduction is neglectable and state-based energy demand as well as production constraints are considered. The authors advance their methodological approach to estimate DR capability of production systems (Sun and Li (2014), described in detail, including background on MDP, within the following sections).

- Also formulating a hybrid flow shop scheduling problem, Luo et al. (2013) include multiple objectives in their model. The two objectives, makespan and electricity demand, are integrated into a formulation which considers TOU prices. The authors propose a multi-objective ant colony optimization to solve their quantitative model. Different scenarios are tested, i.e., the impact of different TOU rate schedules.
- A stochastic model which also utilizes intermediate storage determines the time a production machine can be shut down to reduce total energy demand while maintaining throughput, described in Sun and Li (2013).
- A quantitative approach for reducing energy demand per product and identifying a so-called *Energy Profit Bottleneck* is suggested in Brundage et al. (2014). The purpose is to identify the machine in a serial production line which demands a large amount of energy while exhibiting a low production count, which results in a profit loss. Production dynamics of a serial production line, random failures, and machine warm-up time are considered to determine machines for switch off to reduce energy demand and cost.
- An optimization model has been developed to reduce production cost and energy demand of machines via machine scheduling under varying electricity prices and state-based energy demand in Shrouf et al. (2014).

In addition to above overview, a recently published comprehensive overview on existing methods and research regarding manufacturing scheduling under sustainability goals can be found in Giret et al. (2015). The authors evaluate numerous existing methods, including their differences in inputs and outputs and their utilized scheduling methods. Results include that most approaches focus on energy as (solely) considered input, and CO_2 as (solely) considered output. In addition, nearly all approaches focus on proactive scheduling rather than reactive (considering online information), or hybrid (on-/offline) methods.

Further, in Grismajer and Seliger (2012), a detailed overview of information requirements to align energy demand of manufacturing systems with energy availability is given. The authors explicitly note that multiple reasons for limited energy/electricity availability exists, such as long-term availability or price changes, TOU-pricing, or immediate direct control by suppliers. Several approaches for aligning energy demand and supply are summarized. A structured overview of stakeholders and related tasks is given. Further, motivational aspects for implementing DR actions are discussed, followed by a holistic overview of proposed implementation options for production planners to align stakeholders. Although the approach is directly related to DR and energy flexibility of manufacturing systems, it is not included within the following detailed evaluation as the sole focus is on information requirements rather than technical solutions.

Above discussion of existing research on energy efficiency aims at providing an overview of existing methods as an initial glance at the field and does not claim to be exhaustive. For further detail and additional information see above and related references. The following section summarizes the current state of research on energy flexibility.

3.3 Classification of Energy Flexibility Centered Approaches

Based on the review of approaches with a focus on DR and energy flexibility, a first classification method based on these approaches' methods and background from general theory (Chap. 2) has been developed. The resulting classification matrix can be found in Fig. 3.2. Two dimensions have been chosen: (1) planning horizon and (2) process system type. The first dimension is based on the managerial planning horizons outlined in Sect. 2.1.2 and Fig. 2.4 (strategic management is explicitly included as an important planning horizon, but no relevant approaches were identified). The second dimension, process system type, is based on the generic system stages/network structure (cf. Sect. 2.1.1 and especially Fig. 2.2, special consideration of cyclical flows was not found in reviewed approaches and is therefore omitted).

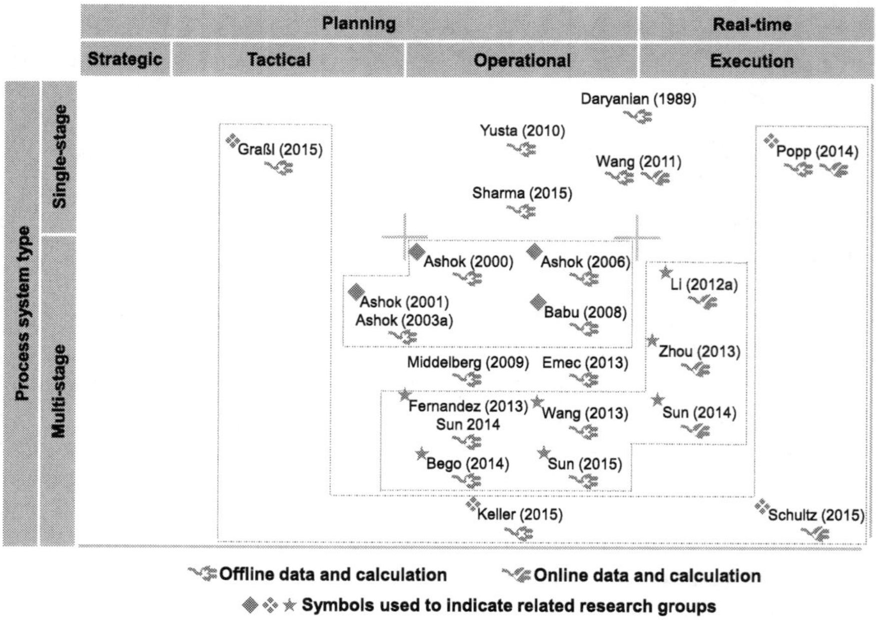

Fig. 3.2 Classification of existing relevant energy flexibility research approaches (only first author given, also for multi-author references, own illustration)

A single-stage structure generally refers to a method which considers one (or more parallel) process(es) which receive input and deliver output without explicitly considering connected system elements, i.e., input/output storage or processes, or systems including process(es) with connected elements that have no (significant) influence on the process(es) behavior (e.g., unlimited storage capacity for incoming/outgoing buffers). In turn, a multi-stage system refers to two or more connected processes which dynamically influence each other, e.g., due to limited intermediate buffer storage capacity. Especially complexity and methodological differences, mainly as a result of inherent dynamics, are different between single- and multi-stage processes. Aside from these two matrix dimensions, a third characteristic is included by assigning each approach a symbol which describes if the approach utilizes offline or online data and calculation. The term offline data is used if an approach uses ex-ante known data (i.e., historic, static or forecast data) and consequently calculates a quantitative result (if any) also offline. Online data and calculation refers to using real-time/online data input for calculating immediate results and potentially feeding results such as production schedules back (e.g., into a manufacturing execution system (MES)). Due to the inherent difference between online and offline data and calculation, most offline approaches are related to (tactical/operational) planning, while online data is required for execution-related approaches. Overall, the position of two approaches relatively within a matrix quadrant has no further classifying meaning. In addition, some approaches might include or could be applicable for more than one planning horizon and are therefore positioned in between quadrants. The same logic is applied for offline/online classification.

Figure 3.3 provides an overview of an initial characterization scheme for existing research approaches. The scheme is designed to provide a basic overview of each approach by summarizing the main goal/purpose of a given approach, indicating which planning horizon(s), basic process structure and evaluation dimensions are addressed and which application example is utilized. *Planning horizon* is split into tactical, operational, and execution time horizon (note that execution is included in the planning horizon category for simplicity). *Process structure* describes general characteristics such as material flow structure, single- or multi-stage systems and if

Goal:

Planning horizon	**Process structure**	**Evaluation**
Tactical	✓ Unidirectional flow	✓ Energy flexibility
✓ Operational	Diverting/converging flow	✓ Operational
✓ Execution	Single-stage	✓ Economic
	✓ Multi-stage	Environmental
	✓ Single process type	
	Various process types	

Application:

Fig. 3.3 Generic overview of basic research characterization scheme with example check marks (own illustration)

only a single (e.g., fixed on/off electricity demand) or various process types (different process behavior, e.g., both on/off and rate-dependent process electricity demand) are included. *Evaluation* summarizes if energy flexibility, operational, economic, and/or environmental indicators are evaluated. A black check mark indicates a direct relevance, while a gray mark indicates an indirect or loose relation/relevance. The scheme aims at providing a standardized overview and as such attempts not to include all detail from a given approach but to give a first structured overview of contents.

3.4 Review of Energy Flexibility Centered Approaches

Structured along the classification matrix provided in picture 3.2, identified existing research approaches are described in the following. The review starts with tactical planning approaches, followed by operational- and execution-related approaches. Within each dimension, approaches are further split into single- and multi-stage process systems.

3.4.1 Tactical Planning

The first method discussed in detail originates from the research project *FOREnergy* (see website for further detail, FOREnergy (2015)). Several publications related to FOREnergy are available and a total of four relevant methods have been identified for detailed review. Table 3.1 summarizes these four approaches and selected related publications. Summarized, among others, theoretical background such as a definition of energy flexibility is provided, an evaluation scheme for energy flexibility proposed and a set of measures and methods to increase energy flexibility via machine control and production planning and control described. First, an evaluation scheme for energy flexibility is summarized in this section as a tactical planning approach. The second, third, and fourth topic (as per Table 3.1) are reviewed within later sections as they address operational and execution time horizons. References in Fig. 3.2 and within later overviews reference the main author or publication within each topic (will be clear from the context) and refer also to related publications listed in the table.

Formally and quantitatively evaluating energy flexibility is regarded as a tactical planning approach within this review as presented evaluation approach includes machine characteristics and measures which are not changed during operational planning but require long(er) term adaption in order to increase energy flexibility. Relevant publications include Reinhart et al. (2012), Reinhart and Graßl (2013), Graßl et al. (2013), Reinhart et al. (2014), Graßl and Reinhart (2014), Graßl (2015). However, selected described measures might be allocated within operational planning, but no operational planning method to determine a specific production plan is given. Further, presented method is centered around single-stage production processes.

Table 3.1 Overview of energy flexibility research related to *FOREnergy*

Topic	Planning horizon	Process system type	Leading author	References
Evaluation of energy flexibility	Tactical	Single-stage	M. Graßl	Reinhart et al. (2012), Reinhart and Graßl (2013), Graßl et al. (2013), Reinhart et al. (2014), Graßl and Reinhart (2014), Graßl (2015)
Energy flexible production planning	Operational	Multi-stage	F. Keller	Popp et al. (2013), Schultz et al. (2014), Liebl and Zäh (2014), Schultz et al. (2015b), Schultz et al. (2015a), Keller et al. (2015), Keller and Reinhart (2015)
Energy flexible machines	Execution	Single-stage	R. Popp	Popp et al. (2013), Popp and Zäh (2014), Liebl et al. (2015)
Energy flexible production control	Execution	Multi-stage	C. Schultz	Schultz et al. (2014), Reinhart and Schultz (2014), Schultz et al. (2015b), Schultz et al. (2015c)

As a starting point, two different types of energy flexibility are described, namely basic and specific energy flexibility (Reinhart et al. 2012). While the first term describes energy flexibility as a function of machine/production system characteristics, the latter includes the influence of a specific production program in an energy flexibility evaluation. The proposed method for basic energy flexibility evaluation is applied for a laser welding station. Further, measures for enabling energy flexibility and their relation to basic production system flexibility types (e.g., throughput- or routing-flexibility) is described (Reinhart and Graßl 2013). In a second step, machine-state models are used for evaluating energy flexibility, which are extended into a Petri-Net based approach (Graßl et al. 2013). A quantitative evaluation method is proposed which yields a dimensionless number reflecting energy flexibility dependent on process variability, flexibility cost and time. A case study is described utilizing a grinding and a laser sintering machine as examples. Proposed evaluation model is extended to include different, additional energy flexibility measures (i.e., delayed processing start, adjustment of processing speed) into the evaluation (Graßl and Reinhart 2014). The resulting evaluation method is discussed in a context to enable electricity price arbitrage as well as participation in reserve markets. A six-step application approach of previously discussed evaluation can be found in Reinhart et al. (2014). These include machine selection, evaluating of machine states, identification of measures, data collection, evaluation, and analysis. A graphitization facility is chosen as application example.

Goal: Evaluate energy flexibility of production machines

Planning horizon	Process structure	Evaluation
✓ Tactical	Unidirectional flow	✓ Energy flexibility
✓ Operational	Diverting/converging flow	✓ Operational
Execution	✓ Single-stage	✓ Economic
	Multi-stage	Environmental
	Single process type	
	✓ Various process types	

Application: Laser welding, grinding, laser sintering, graphitization

Fig. 3.4 Overview of energy flexibility evaluation from Graßl et al. (see Table 3.1 for references, own illustration)

As mentioned before and complementary to above energy flexibility evaluation, within the energy flexibility evaluation, a set of measures has been defined to enable and enact energy flexibility. Among others, these measures include:

- Switch product between machines
- Change production sequence
- Change production start
- Alter shift and staff break times
- Stop/pause processes
- Change process parameters
- Store energy
- Switch energy carrier

However, feasibility and implementation, especially in a dynamic production environment, is not formally addressed.

A summary of described content on energy flexibility evaluation can be found in Graßl (2015). Theoretical background on (general) flexibility in production environments is provided, energy flexibility axioms presented and a dimensionless indicator derived. Further, mentioned six-step application approach is discussed and mentioned case studies described. A summary according to the research evaluation scheme can be found in Fig. 3.4.

Described energy flexibility evaluation approach is used to derive measurements for manufacturing process energy flexibility potential and related load shifting opportunities to profit from volatile energy prices in Feder et al. (2015). However, the specific method or approach how to reschedule loads is not discussed.

3.4.2 Operational Planning

Mathematical planning models for implementing DR actions for single processes have already been proposed in 1989 (Daryanian et al. 1989). For the case of multistage process approaches, several OR-related operational planning methods were identified.

Single-Stage

A linear programming formulation to exploit spot prices via controlling single, energy-intensive processes with connected storage has been proposed by Daryanian et al. (1989). The authors consider only a single process as they note that several industries exist where a single production stage consumes a large share of overall electricity (such as air liquefaction plants or municipal water plants). An overview based on described classification scheme can be found in Fig. 3.5. In order to achieve a reduction in production-related electricity cost, idle capacity is used to reschedule production to times with reduced electricity prices while satisfying (deterministic) demand from final product storage. Planning is performed for one week with hourly varying prices. Known electricity demand of the modeled process is assumed to vary linear with rate of production. Storage capacity is assumed to be limited and associated costs are constant irrespective of utilized capacity while no storage losses occur. The authors note that initial storage and end storage levels have to be held constant to avoid utilization of initial product storage for load shifting purposes as these actions would shift loads to a period outside the planning horizon. No further constraints are included aside from production and storage capacities, especially equipment start-up times have not been considered. The resulting model is solved by a non-simplex algorithm.

An example case study is provided utilizing data from an air compression plant, where a compressor and two liquefier units are scheduled for load shifting purposes. Electricity cost savings are around 10 % for the specific case study, dependent on the weekly spot price schedule used. Sensitivity analysis for several input parameters is performed (impact on percent change in savings), among them changes in initial storage levels and storage and production capacity.

In summary, the presented approach is one of the first optimization methods which explicitly utilize volatile prices in combination with idle production capacity to reduce electricity costs. Due to limited availability of economic computational resources, the resulting model includes only a limited number of restrictions com-

Goal: Reduce electricity for single, storage-type process under spot prices

Planning horizon	**Process structure**	**Evaluation**
✓ Tactical	✓ Unidirectional flow	✓ Energy flexibility
✓ Operational	Diverting/converging flow	Operational
Execution	✓ Single-stage	✓ Economic
	Multi-stage	Environmental
	✓ Single process type	
	Various process types	

Application: Applied for an air compression company

Fig. 3.5 Overview of scheduling approach in Daryanian et al. (1989) (own illustration)

Goal: Max. profit under spot market prices for single machining process

Planning horizon	Process structure	Evaluation
Tactical ✓ Operational Execution	✓ Unidirectional flow Diverting/converging flow ✓ Single-stage Multi-stage ✓ Single process type Various process types	✓ Energy flexibility ✓ Operational ✓ Economic Environmental

Application: Horizontal lathe and example workpiece

Fig. 3.6 Overview of single machine scheduling approach in Yusta et al. (2010) (own illustration)

pared to more recent mathematical optimization models. Nonetheless, the method includes several assumptions which can be regarded as key for DR methods in industry such as constant throughput and impact of storage and production capacity changes. Especially the requirement that initial and final product storage are kept constant to avoid shifting loads outside the planning horizon is important and not followed thoroughly within younger research approaches.

The second single process approach focuses on detailed technical modeling of a machining process (Yusta et al. (2010), corresponding overview in Fig. 3.6). The overall objective is to maximize profit through flexible production scheduling of a single machine under spot market prices. First, a technical model of a machining process to determine production cost is proposed. For example, required power is calculated using electromechanical machine efficiency, cutting speed, depth of cut, feed and process specific energy as input. Further, tool replacement cost and time is included as well as operator costs and electricity price. The resulting expression for unit production costs is nonlinear dependent on cutting speed. A fixed sales price for the product is assumed. The authors then obtain a profit optimization problem with the cutting speed as free variable (and thus number of parts to be produced). With hourly varying electricity prices, determining optimal cutting speed and thus number of parts produced in a given hour extends the problem to be applicable in an electricity spot market scenario. If required, minimum (daily) production targets can be included as constraint. As a result, with hourly changing prices, a profit-maximizing hourly production schedule is obtained.

A horizontal lathe and example workpiece is used to demonstrate the results of the model. Further, 40 different scenarios are used to test the influence of minimum daily production requirements and of electricity price variations. Findings include that relative profit compared to no optimization increases with larger electricity price fluctuations. In addition, higher minimum production requirements reduce profit increase as less scheduling flexibility is available.

Overall, the proposed method includes several technical parameters into a profit-maximizing optimization problem. As such, the specific modeling makes the approach only applicable for machining processes, and, depending on individual process characteristics, the model might need to be significantly reconfigured. Fur-

Goal: Reduce carbon emissions through scheduling and RE generation

Planning horizon	Process structure	Evaluation
Tactical	✓ Unidirectional flow	✓ Energy flexibility
✓ Operational	Diverting/converging flow	✓ Operational
✓ Execution	✓ Single-stage	Economic
	Multi-stage	✓ Environmental
	✓ Single process type	
	Various process types	

Application: Generic example with five periods and twenty tasks

Fig. 3.7 Overview of machine scheduling approach to increase renewable electricity demand presented in Wang et al. (2011) (own illustration)

ther, no product storage and connected system constraints are considered. As a result, the model is only applicable for a single process type and if connected/dependent processes and other systems (e.g., auxiliary systems) are not influenced (and vice versa).

In Wang et al. (2011), the authors propose a method to schedule a manufacturing machine to achieve lower carbon emissions through utilization of renewable electricity (overview in Fig. 3.7). A machine is considered which can process several different tasks, while each task exhibits a different electricity demand during processing. A battery is connected to a renewable energy system (solar and wind) and charged via the RE system. The machine can demand electricity from the grid or battery, while battery (and thus renewable electricity) is always preferred. The planning horizon is divided into different periods, during which renewable energy generation is forecast. All tasks need to be completed during the planning horizon. The main goal is to minimize emissions from renewable and grid electricity demand, direct production emissions as well as additionally assigned (i.e., maintenance) emissions. Tasks are flexibly scheduled and demand a given amount of (renewable and grid) electricity. As renewable electricity has lower emissions than grid electricity in proposed model, emissions can be reduced via utilizing a large share of renewable electricity. However, all tasks need to be completed within the planning horizon, resulting in dependencies between planning/production and emission requirements. The method exhibits characteristics of an execution method which (partly) uses online data as a production planner is included within the application flow who can reapply the scheduling algorithm dependent upon realized and expected (carbon emission) results. No uncertainties and disturbances such as machine failure or operator absence are considered.

A mixed integer programming model to solve described problem is proposed. The authors note that computational complexity is increasing with constraints and especially number of tasks and periods and therefore propose a so-called *low-carbon production scheduling algorithm*. The algorithm ranks production tasks according to their energy demand and successively includes tasks into periods, dependent upon available renewable electricity and battery capacity. The effectiveness of the algorithm is demonstrated using an example application with five periods and twenty tasks.

Overall, the presented method achieves to reduce carbon emissions through flexible production scheduling which aims at increasing utilization of renewable electricity generation. Different tasks with differing electricity demand can be included in the method. However, only a single machine is considered, and dependencies between material flow and especially sequencing of tasks is not considered.

The last approach discussed in the single-stage operational category can be found in Sharma et al. (2015). The authors aim at econological (economic and ecological) efficient scheduling of a manufacturing job shop with multiple machines and multiple parts under a TOU-tariff. Main influencing variables are the operating speed of machines (which can be independently adjusted), and the number of speed changes that machines are allowed to perform in a given period. The authors explicitly focus on discrete manufacturing as they note that load-tracking (and optimization) in discrete manufacturing is believed to be more (technical) complicated than in continuous plants. A two-stage manufacturing setup is considered, in which a set of machines process a part (first stage) and a single-machine second stage assembles the product from processed parts. As all machines are mainly independently scheduled (material flow dynamics are not explicitly considered except that the last stage can only assemble a part when all subparts from the first stage are available), the approach is considered a single-stage approach.

A mathematical model is proposed (based on a two-stage job shop scheduling problem), combined with a multi-objective metaheuristic optimization. From an economic perspective, energy demand and peak demand and their corresponding costs are considered. Total energy (electricity) demand is included as ecological factor. Machines are modeled using detailed energy demand profiles. Further, energy demand is assumed to vary depending on operating speed. In order to limit complexity, only a (preselected) number of operating speed changes is allowed. Different objective functions can be used, i.e., ecological, economic and a combination of both, econological objectives. An aggregate quality index method combined with simulated annealing is applied to obtain a metaheuristic for proposed multi-objective optimization problem. Discrete-event simulation is then implemented to obtain results (electricity load) for a given calculated schedule. An example manufacturing setup and four different operations (milling, drilling, turning, and welding) as well as assembly are part of an example application case.

The standardized overview of discussed approach can be found in Fig. 3.8. In summary, the authors achieve to use energy flexibility, offered by speed changes of manufacturing processes, to simultaneously optimize economic and ecological parameters under a TOU-tariff. Different machine types can be implemented, and different methods (scheduling, simulation and metaheuristic optimization) are combined to yield a (computational) efficient approach. However, inherent material flow dynamics are only briefly discussed, i.e., a two-stage scheduling problem without explicitly considering intermediate buffer capacities cannot be easily extended to a multi-stage problem with limited intermediate storage. Further, as the authors note, (stochastic) availability of machines is not considered.

Goal: Econological efficient manufacturing scheduling under TOU tariff

Planning horizon	**Process structure**	**Evaluation**
Tactical	✓ Unidirectional flow	✓ Energy flexibility
✓ Operational	Diverting/converging flow	✓ Operational
Execution	✓ Single-stage	✓ Economic
	✓ Multi-stage	✓ Environmental
	Single process type	
	✓ Various process types	

Application: Example job-shop scheduling for manufacturing a product

Fig. 3.8 Overview of energy-efficient scheduling approach presented in Sharma et al. (2015) (own illustration)

Multi-stage

The first set of methods and approaches described within this subsection have been developed by S. Ashok et al. Approaches originate from challenges faced in the power sector in India, therefore application examples are also from Indian companies. India experiences a strong surge in energy, especially electricity. As a result, energy and peak power shortages are frequent and require mitigation strategies (Central Electricity Authority 2014). One strategy includes widespread introduction of TOU rates to provide the economic incentive for industrial load management (LM). An overview of approaches developed by or with the participation of Ashok can be found in Table 3.2. Although some approaches are centered around continuous process plants, their methodological logic (e.g., detailed process characteristics) and explicit inclusion of storage capacities for load shifting constitutes the rationale for conducting a detailed research review.

Starting in 2000, Ashok and Banerjee propose a load shifting algorithm to reschedule industrial loads to reduce electricity cost under TOU rates (overview in Fig. 3.9, Ashok and Banerjee (2000)). The authors consider interlocked processes and related product storage constraints. The load shifting algorithm is verbally described and aims at rescheduling loads during a single day to avoid exceeding a predefined maxi-

Goal: Reschedule processes to reduce peak electr. demand and charges

Planning horizon	**Process structure**	**Evaluation**
Tactical	✓ Unidirectional flow	✓ Energy flexibility
✓ Operational	Diverting/converging flow	✓ Operational
Execution	Single-stage	✓ Economic
	✓ Multi-stage	Environmental
	✓ Single process type	
	Various process types	

Application: Fertilizer plant (chemical industry)

Fig. 3.9 Overview of load shifting algorithm method in Ashok and Banerjee (2000) (own illustration)

Table 3.2 Overview of energy flexibility related publications from Ashok et al.

	Ashok and Banerjee (2000)	Ashok and Banerjee (2001)	Ashok (2006)	Babu and Ashok (2008)
Title	Load management applications for the industrial sector	An optimization mode for industrial load management	Peak-load management in steel plants	Peak Load Management in Electrolytic Process Industries
Goal	Reduce elec. cost under TOU rates	Reduce elec. cost under TOU rates, consider LM costs	Reduce elec. cost under TOU rates, consider LM costs, for batch-type processes	Reduce elec. cost under TOU rates, consider LM costs, for continuous/nonlinear processes
Approach	Load shifting algorithm	Integer linear programming (ILP)	Integer linear programming (ILP)	Mixed integer nonlinear programming (MINLP)
Structure	Linked processes, storage constraints, const. throughput	Discrete time, production sequence, operational and electricity demand constraints, const. throughput	Discrete time, detailed electricity load profiles, production sequence, several constraints, const. throughput	Discrete time, nonlinear electr. demand, cont. controllable processes, production sequence, several constraints, const. throughput
Application	Fertilizer plant in India	Flour mill in India; chilled water storage (Ashok and Banerjee 2003a)	Mini steel plants in India	Electrolytic industries (example: caustic-chlorine plant)

mum demand target while maintaining production throughput. If maximum demand is exceeded for a given time period, processes are selected according to their lowest electricity demand and rescheduled to a period where maximum demand is not exceeded. Rescheduling takes storage and interlocked process constraints into consideration, and excludes processes which cannot be rescheduled (such as lighting). New electricity demand and costs are calculated and the procedure repeated until minimum costs for the day are achieved. The method is applied for a fertilizer plant in India. Results include an approx. 4 % reduction in peak demand and approx. 3 % cost reduction, while throughput is maintained as a result of mentioned constraints.

In 2001, the authors publish another optimization approach for industrial LM which utilizes integer linear programming (Ashok and Banerjee (2001), overview in Fig. 3.10). Loads are classified into controllable and noncontrollable (fixed time) loads. Further, controllable loads are divided into process independent, process interlocked, storage constraint and sequential loads (Fig. 3.11). The overall objective is

Goal: Min. energy and load mgmt. costs under TOU rates via scheduling		
Planning horizon	**Process structure**	**Evaluation**
✓ Tactical ✓ Operational Execution	✓ Unidirectional flow ✓ Diverting/converging flow	✓ Energy flexibility ✓ Operational ✓ Economic Environmental
	✓ Single-stage Multi-stage	
	✓ Single process type Various process types	
Application: Flour mill and air conditioning plant of office complex		

Fig. 3.10 Overview of DR approach in Ashok and Banerjee (2001), extended in Ashok and Banerjee (2003a) (own illustration)

to minimize costs under TOU tariffs, which consists of energy costs and additional costs incurred by LM, i.e., labor or capital cost. Maximum demand charges are not explicitly included. The planning horizon is split into a number of discrete time intervals, i.e., quarter to full hours. Decision variables are binary variables which indicate if a given machine is switched on or off in a discrete time interval. Considered constraints are, among others, a minimum output of final products over total time (equivalent to constant throughput), limited product storage capacity, operational constraints such as production sequence and maximum electricity demand. Output of the optimization model is an on/off schedule for all machines and discrete time intervals. The method is exemplarily applied at a flour mill in India. Different tariff structures are tested and various improvement levers employed. These include operation of a third shift (additional labor cost compensated by electricity cost savings) and increasing intermediate product storage capacity. Specific (cost saving) results are strongly dependent on tariff structures and chosen improvement method, but indicate significant saving opportunities. From a methodological point of view, the approach demonstrates that flexibility of input parameters and structure (tariffs, system structure) is required to enable comparison of different improvement options.

In summary, the proposed model provides a tool for industrial LM which combines dynamic system dependencies with several improvement levers (operational and technological). Further, an economic evaluation is performed and the result is

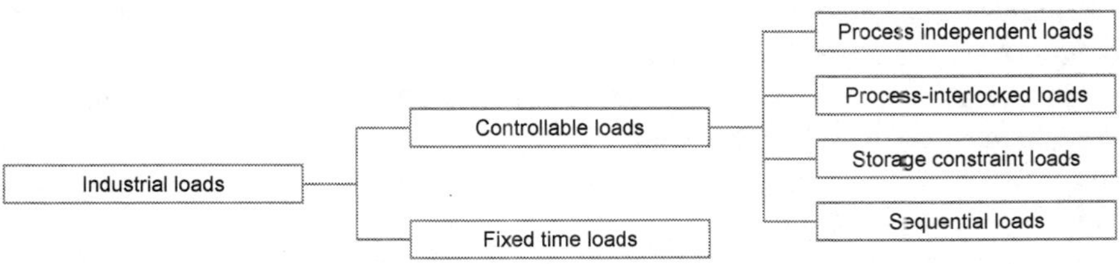

Fig. 3.11 Classification of industrial loads for load management purposes according to Ashok and Banerjee (2001) (own illustration)

optimal under given assumptions. However, discrete time representation and system dependencies (number of connected processes) can result in a high computational complexity if longer time horizons and complex structures are modeled. Further, additional inventory holding costs are not considered. As a minor remark, the proposed flour mill example yields an improvement for a third shift operation as labor costs are low compared to electricity cost (additional shift operation costs approx. 7 % of total electricity costs per day).

Ashok and Banerjee apply their 2001 approach to determine optimal cooled water storage and operating strategy for air conditioning plants (Ashok and Banerjee 2003a). The AC plant is modeled as a set of interlocked processes. Chilled water storage tank capacity and related capital cost are compared to saving opportunities through LM under a given cooling load profile and TOU rates. The resulting nonlinear formulation yields (a) the optimal chilled water storage capacity (and subsequently annualized expenditures) and (b) an optimal operating strategy. A case study is performed for a typical office complex in Mumbai. Results indicate that cost saving opportunities exist, but sensitivity analysis (discount rate for investment cost) reveals that cost saving opportunities and thus LM actions are strongly dependent on individual input parameters (i.e., discount rate). As a continuation/application of Ashok and Banerjee (2001), the approach described in this paragraph will be treated as a single approach together with Ashok and Banerjee (2001), referenced by the first approach. Due to the consideration of different intermediate (product) storage sizes (both in their 2000 and 2001 publication) and related investment, the approach can partly be classified as a tactical planning approach.

In 2006, the previously described approaches from Ashok and Banerjee are further advanced into a model which is based on integer programming for minimizing electricity costs of batch-type plants under TOU-rates (which are enacted to alleviate upcoming peak demand deficits in growing economies such as India) (Ashok 2006). The objective function considers cost from energy demand (unit currency/kWh), peak load demand (unit currency/kW) and additional LM costs (i.e., capital cost for required installations, not further classified). Further, within a case study for a mini steel plant, additional labor costs for night shifts are incorporated. Start times of batch-operated equipment are decision variables. Constraints include minimum production requirements (for multiple products), raw material availability, storage space limitations, and a specified production sequence which products follow. Further, nonproductive time of equipment due to maintenance is considered. The planning horizon is split into a (not further detailed) number of intervals. Load cycles of main equipment are recorded with a 5 min resolution, auxiliary equipment is characterized by an average daily electricity demand value. Solving the described OR problem yields an optimal operating schedule to minimize operating (energy and labor) costs under TOU-tariffs, considering operational constraints. The corresponding overview can be found in Fig. 3.12.

The proposed model successfully integrates different tariff structures into production planning for batch-type plants. However, due to required fixed time steps, complexity and computation time increases if longer planning horizons (i.e., weekly changing tariff structures) or higher temporal resolution is required. Further, the

Goal: Load management under TOU tariff to minimize operating costs

Planning horizon	Process structure	Evaluation
Tactical	✓ Unidirectional flow	✓ Energy flexibility
✓ Operational	✓ Diverting/converging flow	✓ Operational
Execution	Single-stage	✓ Economic
	✓ Multi-stage	Environmental
	Single process type	
	✓ Various process types	

Application: Applied for mini steel plants in India

Fig. 3.12 Overview of load scheduling approach in Ashok (2006) (own illustration)

proposed method considers only one shared storage location and a fixed production sequence. Additional costs incurred by LM are included, but not further differentiated and not dependent on dynamic actions or system state, i.e., inventory holding cost. The presented case study assumes 55 % capacity utilization, which results in approx. 3 % energy demand reduction and approx 5 % electricity cost savings. The authors acknowledge that with increasing capacity utilization effect on peak demand diminishes. A clear advantage of the approach is its applicability to a variety of different industries as it can incorporate batch-type processes and continuous-type processes.

A further refinement of previously described LM approaches can be found in Babu and Ashok (2008) (overview in Fig. 3.13). A MINLP model is developed to schedule continuous processes under TOU tariffs. The total production process is separated into different subprocesses, which are all individually controlled. Each process can be operated with a continuous load factor. Resulting electricity demand, dependent on the load factor, is nonlinear (quadratic fit from observed data). Further, different TOU rates and peak demand restrictions are included as well as operational constraints. A discrete time representation is used, the overall goal is to minimize electricity cost consisting of energy demand and maximum demand charges. Constraints are similar to previous approaches and include mandatory production targets, storage space availability, production sequence and limited electricity availability. As a special characteristic of continuous/process industry plants, minimum raw material flows and related impurities are also considered. A special challenge is caused by the computational complexity of the resulting MINLP problem. However, the authors note that if up-to-date analytic solution techniques are used (modified branch and bound), an optimal solution can be obtained (computation time is not further discussed). The approach is tested for data from a caustic-chlorine plant. Results include 3.9 % saving in electricity costs and 19 % peak demand reduction. As the approach is a general one, the authors note that it can be applied to a variety of process industries. As other electrolytic processes, i.e., extraction of aluminum or zinc, are also characterized by a large share of electricity costs compared to total operating costs, significant (national) cost reduction, energy savings and, from a utility's perspective,

Goal: Min. electricity cost/max. demand under TOU rates

Planning horizon	**Process structure**	**Evaluation**
Tactical	Unidirectional flow	✓ Energy flexibility
✓ Operational	✓ Diverting/converging flow	✓ Operational
Execution	Single-stage	✓ Economic
	✓ Multi-stage	Environmental
	✓ Single process type	
	Various process types	

Application: Applied for caustic-chlorine plant in India

Fig. 3.13 Overview of DR approach in Babu and Ashok (2008) (own illustration)

peak demand reduction might be obtained through widespread deployment of the approach.

Additional approaches with a contribution from Ashok and related to above publications include an optimization model for industrial co-generation (Ashok and Banerjee 2003b). The approach optimizes decentralized electricity generation by industry to minimize cost under TOU tariffs. Further, a nonlinear programming approach is proposed, which utilizes decentralized renewable energy generation for industrial LM purposes (Babu and Ashok 2009). A case study for 22 industries is performed to demonstrate the application and potential impact of the approach. Previously described MINLP formulation (from Babu and Ashok (2008)) has been extended by a graphical user interface (GUI) and a method for implementation using programmable logic controllers (PLCs) in Ashok (2010) and Mohan and Ashok (2011). In general, considered industries are batch-type and continuous industries. The previously described optimization model (with adjustments) is visualized and described using a flowchart and then implemented in a commercially available solver (LINGO). A GUI is developed to enable configuration and parameterization of the model. The model is then solved according to given input, and the resulting production schedule visualized. Further, the schedule is transformed into control signals, which can be interpreted by PLCs. As such, the approach (Babu and Ashok 2008) could be implemented into an automated industrial environment, where widespread use of PLCs is common. Two additional application examples are provided: optimization and control of a steel rolling mill (in Ashok (2010)) and a chemical plant which produces titanium dioxide from ilmenite (in Mohan and Ashok (2011)).

In 2009, Middelberg et al. propose a control model for load shifting under TOU rates (the corresponding overview can be found in Fig. 3.14, Middelberg et al. (2009)). The provided model is a general binary integer programming approach which has an objective function that considers electricity demand of equipment (binary switch between on and off) and changing electricity rates (TOU). A discretized time formulation is utilized. Further, a general vector function for constraints is given without any specific formulation, which, overall, results in a generalized binary integer optimization model. The model is then applied to a South African colliery. In summary, conveyor operating schedules can be optimized to fill product stockpiles or silos

Goal: Min. electr. cost under TOU rates by rescheduling equipment

Planning horizon	Process structure	Evaluation
Tactical	✓ Unidirectional flow	✓ Energy flexibility
✓ Operational	Diverting/converging flow	✓ Operational
Execution	Single-stage	✓ Economic
	✓ Multi-stage	Environmental
	✓ Single process type	
	Various process types	

Application: South African colliery (management of conveyors)

Fig. 3.14 Overview of conveyor scheduling approach in Middelberg et al. (2009) (own illustration)

during off-peak electricity periods and switch conveyors during on-peak periods off. The case study optimization problem includes observed TOU rates from Eskom (South African utility). The conveyor belt system is described and only two conveyor belts are chosen to be included in the LM approach due to product storage availability, individual controllability, and share of energy demand contribution. The second conveyor feeds a silo for loading trains, which arrive at predefined, irregular times. The resulting optimization problem is solved using *Simulink*. Results include 49 % electricity cost savings. In summary, the proposed method is a rather general optimization problem and leaves the task of defining appropriate constraints to the user. These include, for example, additional economic modeling to include inventory holding costs. Further, the proposed method increases significantly in computational complexity if a more detailed temporal resolution is required or more controllable equipment is included.

A mixed integer linear programming model to schedule manufacturing processes under hourly varying electricity prices is proposed in Emec et al. (2013) (overview in Fig. 3.15). The authors consider a sequential manufacturing line with several processing steps and intermediate buffer capacities. The formulated objective function considers time-dependent (known) electricity prices and machine-state dependent energy

Goal: Reduce electricity cost under RTP through scheduling manuf. line

Planning horizon	Process structure	Evaluation
Tactical	✓ Unidirectional flow	✓ Energy flexibility
✓ Operational	Diverting/converging flow	Operational
Execution	Single-stage	✓ Economic
	✓ Multi-stage	Environmental
	✓ Single process type	
	Various process types	

Application: Example includes line with 8 machines and 20 orders

Fig. 3.15 Overview of RTP manufacturing line scheduling approach in Emec et al. (2013) (RTP: real-time pricing, own illustration)

Goal: Min. electr./inventory cost by shifting production via added inventory

Planning horizon	**Process structure**	**Evaluation**
Tactical	✓ Unidirectional flow	✓ Energy flexibility
✓ Operational	Diverting/converging flow	✓ Operational
Execution	Single-stage	✓ Economic
	✓ Multi-stage	Environmental
	✓ Single process type	
	Various process types	

Application: Automotive assembly line

Fig. 3.16 Overview of *Just-for-Peak* buffer inventory approach in Fernandez et al. (2013) and Sun et al. (2014) (own illustration)

demand. Two machine states, processing and waiting, are decision variables, while a third state, idle, is assumed when a certain time period (e.g., 15 min) in waiting has been passed. The model's planning horizon is one day (24 h). Constraints within the model include mandatory production requirements within the planning horizon, maximum electricity demand and predefined production sequences of products. The resulting production schedule minimizes electricity costs considering hourly spot market wholesale prices, e.g., from EEX. A time step of 5 min is chosen to manage computational complexity, which is addressed as a limiting factor within the discussion of the approach. A working program prototype is presented and applied for an example case. Different scenarios, for example, no energy-related scheduling, constant prices, and increased starting inventory, are evaluated. Results include cost savings of more than 70 % compared to a scenario with fixed prices (and thus no energy-related scheduling).

As an operational planning (scheduling) method, the proposed methodology extends previously discussed production planning methods by explicitly considering machine states and related energy demand. However, overall system energy demand is not evaluated, and cost savings are related to wholesale pricing and neglect further electricity-related cost such as transmission and peak demand charges. Operational KPIs, e.g., impact on throughput time, are not reported. A validation through implementation into, e.g., a simulation environment is also not mentioned. As the authors note, a discretization into 5 min intervals omits detail within power demand patterns.

A second set of operational planning and execution energy flexibility approaches originates from L. Li from the University of Illinois, Chicago. The first discussed method aims at utilizing *Just-for-Peak* inventory in connected manufacturing lines for load shifting purposes and has been developed by Fernandez, Li and Sun (overview in Fig. 3.16, Fernandez et al. (2013), with an extension in Sun et al. (2014)). The objective of proposed approach is to reduce peak demand of manufacturing systems in order to avoid electricity grid extension requirements. The authors consider a sequential manufacturing line with several processes which are connected via buffers with limited capacity. Production is scheduled for a known future period, which consists of an off-peak period with a charge for total energy demanded (i.e., USD/kWh), followed by an on-peak period characterized by costs for total energy demand and peak demand (i.e., USD/kW). Additional buffer space is deployed between machines to build inventory during off-peak periods and consume inventory during on-peak period if the adjacent upstream machine is switched off (constant system throughput requirement). Machines are characterized by stochastic failures and repair times, a given cycle time and power demand during production.

A nonlinear integer programming (NIP) approach is used to minimize the total cost per unit time. Electricity demand charges (total energy and peak charge) and inventory holding costs for accumulated additional inventory are part of the objective function. Additional inventory is assumed to be able to accumulate with a linear rate (additional upstream production capacity) and is consumed linearly if the upstream machine is switched off during the peak period. Four constraints are formulated: (1) a certain minimum electricity demand reduction requirement needs to be achieved, (2) enough additional intermediate products need to be available, (3) buffer capacity cannot be exceeded and (4) the last machine is never switched off (constant throughput). The resulting problem is solved using commercially available software (*GAMS*), which utilizes an extended solver based on the standard branch and bound method. The authors infer that computation time for up to 100 machines should still be acceptable as no real-time solution is required. Simulation is used to evaluate system behavior when results from the NIP problem solution are enacted (switching machines off during peak period). A case study demonstrates the application of the method (part of an automotive assembly line with seven machines and six buffers). Aside from successfully implementing the approach, results include an approx. 15–20 % reduction of both power demand (kW) and costs.

The proposed approach explicitly utilizes embodied energy in intermediate inventories to reshape electricity demand of a manufacturing system. Being a NIP approach, total charges can be optimized (minimized), but with the downside of preventing a real-time solution for complex problems due to computation time. Further, the specific formulation of the objective function can only consider a planning horizon characterized by an off-peak and then on-peak period. However, the solution is independent from system start and end state as only additional inventory is utilized for demand reduction.

The previously discussed approach is advanced in Sun et al. (2014) by relaxing the throughput constraint (for further discussions, Sun et al. (2014) and Fernandez et al. (2013) will be considered as a single approach summarized under Fernandez et al. (2013)). The overall methodological approach remains the same, but additionally built buffer inventory is not required to ensure constant system throughput. Consequently, an economic evaluation of throughput loss is included to find an economic optimum. In summary, the decision for each production machine is to either produce/run through the DR event, being switched off or, if turned off at the beginning of the DR event, being turned on again during the DR event if required to increase/maintain throughput. The authors use the same example case from the initial study and demonstrate that a further improvement of peak energy demand (initial reduction 18.1 %, increased to 26.8 %) and total cost (initial reduction 16.7 %, increased to 25.3 %) is feasible.

Goal: Min. electricity demand or cost under TOU rates of linked processes

Planning horizon	Process structure	Evaluation
Tactical	✓ Unidirectional flow	✓ Energy flexibility
✓ Operational	Diverting/converging flow	✓ Operational
Execution	Single-stage	✓ Economic
	✓ Multi-stage	Environmental
	✓ Single process type	
	Various process types	

Application: Generic example case with three machines and two buffers

Fig. 3.17 Overview of electricity demand and cost reduction approach in Wang and Li (2013) (own illustration)

The second approach from Chicago describes a detailed mathematical model for LM purposes in Wang and Li (2013). A zero-one linear programming approach is used to schedule manufacturing machines with connected buffers under TOU rates to either minimize total electricity demand or electricity cost (overview in Fig. 3.17). A discrete time formulation is used to model the evolution of a sequential production line. System evolution is modeled in a generalized form $\mathbf{X}(t) = G[\mathbf{X}(t-1)]$, where $\mathbf{X}(t)$ describes state probabilities of the system at t, which are obtained using the system states $\mathbf{X}(t-1)$ and the function $G(\cdot)$, which denotes the dynamic transition of the system. Further, stochastic failures of machines and additional indicators such as work-in-progress (WIP), system output and power demand are considered. Electricity demand charges consist of energy and peak demand charges. The generalized system state formulation is used to define an optimization problem. The authors choose to investigate two optimization goals, namely minimizing total electricity demand and electricity cost. The resulting optimization problem is a zero-one nonlinear programming problem (ZONLP), which is a subclass of the MINLP problem class. Consequently, only approximate methods are available to obtain a solution. Within the discussed approach, the authors choose to utilize particle swarm optimization (PSO) to determine a production schedule. A generic case study is used to demonstrate the method: a process chain with three machines and two buffers is modeled, parameters are freely chosen. The case study shows that a solution can be found within reasonable time utilizing a conventional desktop computer. Further, time series for the machines' control signals are visualized as well as additional indicators such as WIP, system production rates, energy demand and energy cost. Sensitivity analysis is performed to investigate the influence of machine reliability and buffer capacity on the objective. In summary, the proposed approach achieves to provide detailed insight into the (mathematical) complexity of manufacturing system scheduling under time-varying electricity costs. Sensitivity analysis demonstrates that changing system parameters can have significant impact on the obtained near-optimal solution.

Related to previously described approaches from Li and Sun, a MINLP approach has been proposed to minimize total electricity cost and penalties for lost production

Goal: Min. electr. and production non-fulfillment cost under CPP tariffs		
Planning horizon	**Process structure**	**Evaluation**
Tactical	✓ Unidirectional flow	✓ Energy flexibility
✓ Operational	Diverting/converging flow	✓ Operational
Execution	Single-stage	✓ Economic
	✓ Multi-stage	Environmental
	✓ Single process type	
	Various process types	
Application: Serial production line from industry (not further detailed)		

Fig. 3.18 Overview of manufacturing system scheduling under critical peak pricing (CPP) tariffs presented in Bego et al. (2014) (own illustration)

under CPP tariffs (Bego et al. 2014). The corresponding overview can be found in Fig. 3.18.

CPP rate structure consists of the CPP interval (extreme charges) and an otherwise TOU tariff with on- and off-peak periods. Further, charges for reservation capacity are included. In the described example, reservation capacity can be bought at a fixed charge and determines how much electricity demand is excluded from extreme charges during CPP events. Consequently, the reservation capacity level and production scheduling are decision variables in the MINLP formulation to minimize electricity and demand nonfulfillment cost. Total time for a given billing cycle is divided into time intervals with equal length. Electricity demand is calculated dependent on a given production schedule (which machines are on or off during a time interval) and, depending on rate structure (TOU rates, reservation capacity, etc.), total costs are calculated. Random failures of machines are not included (which leads to an overestimation of electricity demand as demand is lower during a failure event). Production during a given time interval is calculated depending on machine efficiency and schedule. A set target production is required and formulated as a constraint in the optimization model. Further constraints are introduced, for example, buffer capacity constraints and inventory balance.

Being a MINLP formulation, computational complexity is a central concern. Therefore, the authors use an approximate technique which divides the global problem into several subproblems. The total time interval is separated into subintervals (each containing multiple time step intervals). The resulting subproblem is then solved and the solution set as given input for the next subinterval. Commercially available software packages are used to solve the MINLP formulation.

A case study is used to exemplarily apply the approach. A sequential production line with five machines and four buffers is modeled, and real CPP rates from San Diego Gas and Electric are included. Total model time is four weeks, which are separated into four subproblems of one week duration, time step length is one hour. After solving the problem and thus obtaining an optimized production schedule, a discrete-event simulation tool is used to determine total system throughput. Optimized reservation capacity, demand nonfulfillment and total costs as well as detailed cost components are determined. For comparison, a so-called *simple method* (zero reservation capacity) and *expert rule* (reservation capacity is set to half of maximum demand) are introduced. Results show that cost savings between approx. 31 % (simple rule) and approx. 11 % are obtained by the proposed method.

In summary, the presented approach provides a model for minimizing selected costs under CPP for manufacturing enterprises. Especially a long planning horizon (one month) and consideration of economic parameters for throughput loss differentiate this planning/optimization approach from previously presented research. However, stochastic failure or random CPP events are not included. Further, computational complexity, especially for longer time horizons or more complex manufacturing systems, might be a challenge when applying the method to real-life problems.

Another approach from the research group around L. Li advances previously described methods from Bego, Li and Sun by minimizing costs of a manufacturing system with on-site CHP generation through scheduling of manufacturing machines and a CHP plant under TOU rates (Sun et al. 2015). Similar to previously presented methods, the approach considers dynamic interactions of the production machines (connected with intermediate buffers), related dynamic electricity demand and achieves a minimum throughput (overview in Fig. 3.19). The resulting model is a MINLP approach and an approximate solution is obtained via PSO. The optimization goal is to minimize costs, which consist of electricity energy and demand charges under a TOU tariff (on-/off-peak periods) and operating costs for the on-site CHP plant. For each time interval, a production schedule for production machines and operating schedule for the CHP generation needs to be determined (decision variables). The CHP plant and auxiliary boiler is modeled utilizing characteristic values such as natural gas calorific value, electricity to heat ratio, efficiency of the boiler and CHP, set-up costs and fuel costs. Electricity generated from CHP cannot be sold back to the grid. Constraints of the MINLP formulation include material flow balance, minimum target production, minimum on- and off-times and maximum output of the CHP plant. Computational complexity of the problem presents a challenge due to nonlinear equations and discrete variables. Consequently, PSO is used to obtain a near-optimal solution. The presented method is tested using a

Goal: Min. cost under TOU rates by production and CHP plant scheduling

Planning horizon	**Process structure**	**Evaluation**
Tactical	✓ Unidirectional flow	✓ Energy flexibility
✓ Operational	Diverting/converging flow	✓ Operational
Execution	Single-stage	✓ Economic
	✓ Multi-stage	Environmental
	✓ Single process type	
	Various process types	

Application: Fictitious manufacturing system and CHP plant

Fig. 3.19 Overview of combined production and CHP plant scheduling approach in Sun et al. (2015) (CHP: combined heat and power, own illustration)

fictitious manufacturing system with five processes and four buffers, as well as a connected CHP plant and auxiliary boiler located in Chicago (to determine heat requirements). Simulation time is eight hours, which is divided into 15 min intervals. The example case achieves a relative cost saving of 23.14 % compared to a scenario without CHP. In summary, the proposed method extends previously discussed approaches by explicitly including selected auxiliary systems. The approach effectively demonstrates that integrated control of energy demanding processes (i.e., auxiliary services and manufacturing processes) can yield significant opportunities for energy and/or cost savings. However, previously discussed computational complexity (MINLP formulation) provides challenges for larger problems (e.g., increased number of machines, longer/more granular time horizon).

The last approach discussed in the operational planning category is the approach from FOREnergy (cf. Table 3.1, references include Popp et al. (2013), Schultz et al. (2014), Liebl and Zäh (2014), Schultz et al. (2015a, b), Keller et al. (2015), Keller and Reinhart (2015)), the corresponding overview can be found in Fig. 3.20. Summarized, an energy-aware/energy flexible production planning and control method is proposed to align a factory's energy demand to volatile energy supply and/or prices. As a starting point, a modular simulation method to determine energy demand of a factory is proposed in Liebl and Zäh (2014). A multi-level model is proposed, which splits energy demand of different entities into state-based and product-based demand. Proposed concept accounts for differences in data availability and notes that both a detailed (e.g., product and state-based) and a generalized (e.g., cumulative profile for several machines) profile can be integrated. The resulting energy demand forecasting is then used to increase planning accuracy for energy-related production planning. Schultz et al. (2014) include a conceptual overview on how an energy oriented energy production planning and control (called *ePPS*) should be structured. A focus is set on differentiating production planning and production control in relation to different time horizons from different energy markets (i.e., forward contracts, day-ahead and intraday markets). Concluding, an energy-oriented production planning should determine a desired production plan including energy as a planning constraint, for example, utilizing hourly varying prices. The resulting

Goal: Energy-flexible production planning

Planning horizon	**Process structure**	**Evaluation**
✓ Tactical	Unidirectional flow	✓ Energy flexibility
✓ Operational	✓ Diverting/converging flow	Operational
Execution	Single-stage	✓ Economic
	✓ Multi-stage	Environmental
	✓ Single process type	
	Various process types	

Application: Generic process chain (five steps, parallel machines)

Fig. 3.20 Overview of energy flexibility production planning and scheduling approach presented by Keller et al. (see Table 3.1 for references, own illustration)

(hourly) planned energy levels should then be realized/maintained by production control utilizing online information. A conceptual information model for an energy-oriented PPS is presented. No further application or implementation is discussed. Proposed energy-oriented ePPS is further discussed in Schultz et al. (2015b), especially with a focus on economic implications. Threshold cost values for installing flexibility (overcapacity) are discussed. In addition, relevance of on-site generation is mentioned.

Related to previous discussion, a quantitative production planning approach is presented in Keller et al. (2015). An energy-oriented production scheduling for hybrid flow shops is presented. A machine-state dependent energy demand model is used. The discussed manufacturing system includes five steps with parallel machines. A so-called energy limit is introduced, which is dependent upon current and average energy prices. A target function is formulated, which includes total energy demand, energy limit and capacity utilization. The quantitative model is not provided in detail, but a discussion on solution techniques is included. A simulated annealing method is chosen and results as well as computational performance evaluated. In summary, the method achieves a reduction in both energy demand and energy cost. Nonetheless, increasing number of iterations for the heuristic improves results, but at a significant increase in computational cost. For the specific case modeled, computational time reaches from approx. 12–36 h, indicating that application for larger manufacturing systems is challenging.

A further conceptual discussion and extension of energy-oriented production planning can be found in Keller and Reinhart (2015). Characteristics in relation to production planning of on-site/own generation as well as energy storage (batteries) are presented. The relation between energy procurement, production planning and related different planning horizons, energy sources, and costs are highlighted.

After presenting relevant operational planning approaches in the field of energy flexibility, the following section discusses execution-related approaches.

3.4.3 Execution

After discussing relevant tactical and operational planning approaches, this section summarizes execution-related, i.e., direct production control approaches.

Single-Stage

The third topic from FOREnergy is related to enable energy flexibility of single machines without compromising throughput or product quality by controlling peripheral systems with (energy) storage capabilities such as hydraulic pumps or chip conveyors (compare to Table 3.1, references include Popp et al. (2013), Popp and Zäh (2014), Liebl et al. (2015)). Essentially, peripheral systems which are switched on or off dependent on a given state of the machine (i.e., pressure level of a hydraulic system) could be directly controlled and thus timing of energy demand influenced. This type of energy flexibility is enabled through intermediate storage capabilities, for example, a high-pressure tank which can operate within a certain pressure band.

Rather than repressurizing the tank if a certain minimum threshold is met, the authors propose to pressurize the tank dependent upon load shifting goals, i.e., electricity price. For the case of a chip conveyor, load shifting is enabled through buffering swarf within the machine until a predefined level. A discussion of evaluating components' energy demand and identifying suitable components is included. Further, in Liebl et al. (2015), previously discussed energy flexibility evaluation developed by Graßl et al. is extended by including discussed options for load shifting and a differentiation into organizational (i.e., differing of production start) and technical (i.e., shifting a (sub-)process start) load shifting. Evaluation examples include a milling machine, a laser cutting machine and a reflow oven (among others).

Overall, the presented method highlights the opportunity for load shifting via using (energy) buffers within machines without directly affecting production parameters such as throughput or quality. However, no concrete application example and/or method is provided (the authors note that measures are currently implemented at a milling machine). Further, the authors highlight that including (sub-)processes which have an effect on start or end dates of part processing requires consideration of logistical constraints in an interlinked manufacturing system. The corresponding graphical summary can be found in Fig. 3.21.

As indicated in Fig. 3.2, operational planning approaches for single processes might also find their application for execution purposes. The main reason is missing (significant) dynamics and thus nonlinearities and complexities due to (nondeterministic) system evolution. In particular, a single process is characterized by a limited number of states (i.e., on/off, idle, production, failure and related process parameters). Consequently, decision rules for DR actions can be determined for all (discrete or continuous) states of the process (i.e., "in case of DR request, switch process off if a certain number of finished parts is available"). Therefore, existing planning approaches can be used to determine a set of static rules, e.g., summarized in a lookup table, which can be applied to enable energy flexibility of single processes. Another related example are air conditioning (cooling) applications used for DR purposes. Depending on current air characteristics (inside and outside temperature, moisture level, etc.), a static set of rules determines if equipment can be used for LM.

Goal: Energy flexibility of machines through control of indirect processes

Planning horizon	**Process structure**	**Evaluation**
Tactical	Unidirectional flow	✓ Energy flexibility
Operational	Diverting/converging flow	✓ Operational
✓ Execution	✓ Single-stage	Economic
	Multi-stage	Environmental
	Single process type	
	✓ Various process types	

Application: Milling machine, laser cutting, reflow oven and other

Fig. 3.21 Overview of energy flexibility approach via controlling indirect machine processes from Popp et al. (see Table 3.1 for references, own illustration)

Multi-stage

Execution and thus real-time DR and energy flexibility approaches are rarely found compared to planning approaches. Approaches which offer real-time evaluation and/or control of manufacturing systems face several challenges, among them (Li et al. 2012a):

1. **Constant throughput**: Throughput and consequently nonfulfillment of mandatory production targets (customer delivery commitments) and lost margin of sales are of major concern for manufacturing companies. Especially for interlinked systems, impact on system throughput is hard to quantify in real-time. Without operational impact and thereon based economic impact estimation, profit-maximizing companies are unlikely to put throughput for nonquantifiable energy (cost) savings at risk.
2. **Dynamic complexity of manufacturing systems**: Interlinked manufacturing systems can assume a large number of different states, dependent on the system's evolution. Most (real) systems observe stochastic behavior (i.e., due to stochastic failures and downtime of machines), combined with dynamic interaction of system elements (i.e., blocking and starvation of processes). As such, it seems to be impossible to create a set of simple rules what action should be performed dependent on a given system state.
3. **Real-time requirement**: Real-time approaches require data collection, processing and decision-making within a predefined (short) period of time. With regards to data collection, online data sampling and automated collection needs to be available. For real-time decision-making, collected online input data needs to be processed, conclusions drawn and appropriate actions initiated within a predefined time. Consequently, computational requirements need to be known (for all possible types of input combinations) and enough/appropriate resources available to ensure a response within a given time frame.

Consequently, only few approaches can be found in existing literature. The first approach discussed in this subcategory originates from the research group around Li et al. (2012a). An overview can be found in Fig. 3.22. The general objective of the approach is to improve the stability of the (U.S.) power grid by implementing appropriate DR actions exercised by manufacturing companies with interlinked manufacturing systems. A special focus lies on real-time DR, which is required as a response to an immediate and random grid emergency situation such as transmission congestion or generation capacity failure. The considered manufacturing system consists of several connected machines and buffers, online information is used for decision-making. The method aims at achieving a constant throughout, i.e., the sudden DR action ought not to compromise overall system production output. Machines underlie stochastic failures and repair times and average production electricity demand and cycle times are given. Buffers have an initial content and maximum capacity. DR events are modeled as stochastic events. Their beginning time, duration and electricity demand reduction requirement follows a normal distribution.

The methodology's first step is to determine all machine shutdown combinations which result in the required electricity demand reduction. Simulation is used to

Goal: Switch-off machines to accommodate short-term DR event

Planning horizon	Process structure	Evaluation
Tactical Operational ✓ Execution	✓ Unidirectional flow ✓ Diverting/converging flow ✓ Single-stage Multi-stage ✓ Single process type Various process types	✓ Energy flexibility ✓ Operational Economic Environmental

Application: Section of an automotive assembly line

Fig. 3.22 Overview of DR approach in Li et al. (2012a) (own illustration)

determine system throughput for the different shutdown options, which is compared to a throughput baseline (without shutdown) to determine which options are viable to satisfy both electricity demand and throughput targets. The method is tested using data from a section of an automotive assembly line (seven workstations including parallel streams, five buffers, Fig. 3.23). The considered planning horizon is an eight hour shift, for which 60 replications of the simulation model are evaluated. Numerical results are obtained and compared to, e.g., a non-real-time strategy where machines are switched off in descending order of their rated power.

Overall, the proposed methodology achieves the goal of determining if a required electricity demand reduction can be accomplished by a manufacturing system without compromising throughput. Further, the real-time strategy yields superior results compared to a non-real-time strategy. The method considers the current state of the system based on online data and provides an answer in a very short period of time. As a result, a manufacturing company could utilize the method to decide how to respond to a short-term/immediate DR request. However, the authors note that with increasing DR event duration, further evaluation is required to quantify the potential trade-off between throughput loss and DR capability. In addition, the method does not include a long-term analysis of potential impacts from the DR action. For example, long-term effects on throughput and system capability for a second or more frequent DR events would be of interest.

The second execution approach discussed here was published by Zhou and Li (2013), the corresponding overview can be found in Fig. 3.24. The overall goal is to accommodate a short-term DR event by switching machines in a manufacturing line off. The authors consider online production data for bottleneck detection of a

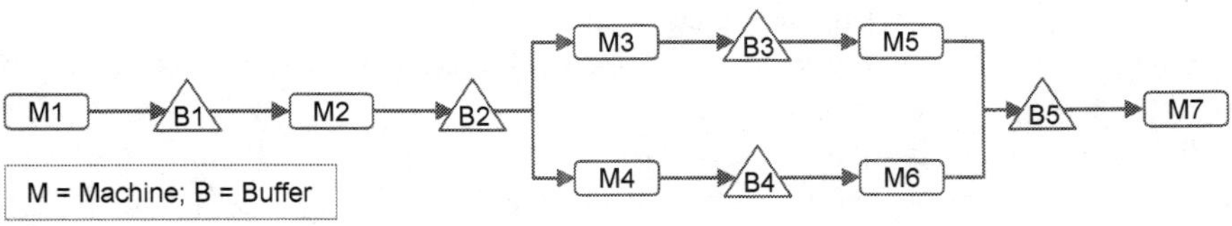

Fig. 3.23 Case study manufacturing system layout used in Li et al. (2012a) and Zhou and Li (2013) (own illustration based on cited references, original copyright 2012, 2013 IEEE)

Goal: Switch-off machines for short-term DR event using bottleneck inform.

Planning horizon	**Process structure**	**Evaluation**
Tactical Operational ✓ Execution	✓ Unidirectional flow ✓ Diverting/converging flow ✓ Single-stage Multi-stage ✓ Single process type Various process types	✓ Energy flexibility ✓ Operational Economic Environmental

Application: Automotive assembly line

Fig. 3.24 Overview of DR approach using bottleneck detection presented in Zhou and Li (2013) (own illustration)

production system. This information is utilized to prevent production losses in case of a DR event by determining the bottleneck of the system to avoid switching off bottleneck machines. The considered production system is an unidirectional flow line with limited intermediate buffer space. Machines are characterized by an individual electricity demand during operation. Further, machines can be down due to stochastic failures. Both machine states (switched off or failure) result in zero electricity demand. Bottleneck detection is based on the assumption that the machine with the highest impact on overall system throughput is the bottleneck machine. As the bottleneck machine cannot be directly identified in a complex system layout due to nonlinearities, the authors utilize the approach presented in Li et al. (2009). The presented method considers blockage and starvation probabilities of machines (dependent on upstream and downstream buffer capacities and inventory) to identify the system's bottleneck machine. The method is advanced in Zhou and Li (2013) to identify a so called *slack point*, which is defined as the machine which has the highest sum of blockage and starvation time of all machines within the system. In order to respond to a given DR event, the DR event is split into time intervals Δt. For each Δt, bottleneck machines are identified and, together with failed machines, excluded from a list containing all machines. This results in a list which includes all machines for potential shutdown. These remaining machines are characterized by their total blockage plus starvation time, and machines are switched off starting with the machine which has the highest total blockage plus starvation time. Additional machines are switched off until electricity demand is equal or lower than required by the DR event. During Δt, total system electricity demand is observed and if demand is higher than required by the DR event, additional machine(s) are switched off. Δt is set to be 15 min as, for the assumed application, maximum electricity demand is averaged over this time interval. Further, the presented method is heuristic and thus an optimal solution is not guaranteed.

To demonstrate the effectiveness of the approach, the authors utilize the same layout and similar system parameters (mean time between failures (MTBF), mean time to repair (MTTR), electricity demand, cycle time) for an automotive assembly line as in Li et al. (2012a) (Fig. 3.23). Time intervals and starting points for DR events

Goal: Estimate potential of manuf. System for event-driven DR event

Planning horizon	Process structure	Evaluation
Tactical Operational ✓ Execution	✓ Unidirectional flow Diverting/converging flow	✓ Energy flexibility ✓ Operational Economic Environmental
	✓ Single-stage Multi-stage	
	✓ Single process type Various process types	

Application: Serial production line from industry

Fig. 3.25 Overview of event-driven DR approach in Sun and Li (2014) (own illustration)

are normally distributed with $N(\mu, \sigma)$, where μ is the mean and σ the standard deviation. Mean time duration for DR events are between one hour and six hours, mean starting point is after two hours. Further, each DR event constitutes a maximum electricity demand limit which cannot be exceeded. Results of the proposed control strategy are compared to a strategy which randomly switches off machines until the required electricity demand reduction is fulfilled. Simulation for an eight-hour shift is used to obtain numerical results. Results indicate that the proposed control strategy achieves, on average, a higher throughput than the simple random control strategy. However, system throughput is jeopardized compared to not responding to a DR event, and throughput losses increase with an increased DR event duration (the specific case study uses one to six hours average DR event duration).

In summary, the approach utilizes real-time system bottleneck information to avoid throughput loss as a DR event consequence. However, constant throughput is not guaranteed and therefore an economic quantification is required (as noted by the authors). Further, no time constants (switch-on/off duration) and only two machine states are considered. No indication is given how machines which are currently processing a part are controllable (immediately stoppable, wait until finished or stopping results in rejected part). Further, difference between initial and final system state (i.e., buffer inventory) is not accounted for. As noted before, being a heuristic approach, an optimal solution is not guaranteed.

Another approach included in this review from Zeyi Sun and Lin Li is also an execution-related approach, the standardized overview can be found in Fig. 3.25 (Sun and Li 2014). The approach focuses on estimating the potential of manufacturing systems to respond to event-driven DR requirements. The considered manufacturing system consists of several directly connected machines and buffers, and the approach aims at providing real-time applicability and achieving constant throughput. A model is proposed which determines control actions for machines and calculates the capability of the manufacturing system for electricity demand reduction. A MDP methodology is used to achieve mentioned targets. The proposed method is based on previous work from the authors which utilizes MDP for energy saving purposes (Li and Sun 2013).

Briefly described, an MDP approach assumes that at a system is in a certain state, in which the decision-maker has the choice between different actions, dependent on the system state. Depending on the chosen action, the system advances into a different state in which the decision maker has (another) set of options available to choose from. For the approach described here, the system state is described by machine operating states, corresponding energy consumption and productivity as well as buffer state. The decision-maker can adopt a power control strategy, which is the action within the MDP framework. Random failures of machines are reflected by uncertain system evolution pathways. The method aims at minimizing electricity demand during a given DR event, which is achieved by determining appropriate control actions. Machine states can change between off, several hibernation states, ready, and full operation state. In order to reflect system dependencies, machines are either the bottleneck of the system or not. Buffers are characterized by their individual fill level (number of parts). Three power control actions for machines are defined: (1) state remains unchanged, (2) switch from ready to hibernation or off (3) switch from hibernation or off to ready. Determination of bottlenecks relies on previously discussed method from Li et al. (2009).

Several constraining assumptions need to be taken into account when applying the method, among them that the DR event is of short duration (10–20 min), equal machine cycle times, and that DR event notification to DR end equals a multiple of cycle times, as the time horizon is divided into time periods equal to (all machines') cycle time. Due to computational complexity, calculating outcomes of all possible states and corresponding actions (full backward recursion) is only feasible for small problems. Consequently, the alternative method used provides an approximate solution (forward method) and is therefore not able to guarantee an optimal solution.

A case study of a serial production line is used to demonstrate the effectiveness of the approach. The serial line consists of five machines and four buffers, and the tested DR event has a duration of 15 min. Applying calculated energy control actions, an approximate 20 % demand reduction while reducing throughput by only 0.2 % is achieved. The authors note that computation time on an average desktop computer is approximately 10 min and therefore the approach qualifies as a real-time approach (assuming, e.g., a 15 min notification period from the utility for a DR event).

In summary, the method achieves to provide a real-time method to estimate potential demand reduction under event-driven DR programs considering dynamic system dependencies. However, several constraints need to be considered (see above). For the case of more complex systems and/or longer time duration of DR events, computational complexity will hinder real-time applicability or reduce the quality of the solution as less iterations within the MDP solution can be calculated. Further, although throughput is held constant through the duration of the DR event, long-term impact on system performance (i.e., due to different beginning and end states of buffer fill levels) should be considered.

The last approach discussed within this review is the forth approach from FOREnergy (Table 3.1, overview in Fig. 3.26), which aims at energy flexible production control. In addition to previously summarized and discussed references Schultz et al. (2014, 2015b), Reinhart and Schultz (2014) and Schultz et al. (2015c) include detail

Goal: Energy-flexible production control		
Planning horizon	**Process structure**	**Evaluation**
Tactical	Unidirectional flow	✓ Energy flexibility
✓ Operational	✓ Diverting/converging flow	✓ Operational
✓ Execution	Single-stage	Economic
	✓ Multi-stage	Environmental
	✓ Single process type	
	Various process types	
Application: Gear box production in a job shop environment		

Fig. 3.26 Overview of energy flexibility production control approach suggested by Schultz et al. (see Table 3.1 for references, own illustration)

on proposed (real-time) control method. In Reinhart and Schultz (2014), requirements for an energy flexible control and their relation to existing energy management systems are formulated. The authors note that energy flexible control tasks are well suited for integration into an (existing) MES. However, little or no functionality in commercial MES with regards to load shifting actions is implemented. In Schultz et al. (2015c), a real-time energy-related order release method is proposed. The method considers a given production and thus energy production schedule and aims at reducing impact of deviations from planned (energy) values. Deviations can be caused, for example, by random machine failures or unexpected changes in energy supply. A two-step order release method is proposed which determines if a job/order in a job shop environment can be rescheduled (i.e., delayed or switched to another machine) without risking to not deliver the respective order on time. If an order waiting in front of a machine can be delayed or switched, potential energy scheduling requirements are evaluated. If electricity demand differs from planned demand (positive or negative), orders are rescheduled or switched between machines to achieve a better fit between planned demand and actual demand. A simulation case study is used to demonstrate the effectiveness of the approach, utilizing a job shop in which different gear boxes are manufactured as example. Results indicate that the presented method achieves an improved fit of actual electricity demand with planned demand compared to a case where no energy flexibility control is enacted. However, throughput cannot be held constant and delivery reliability is slightly reduced, which is noted as an improvement area by the authors. In summary, the presented method achieves an energy flexible production control by utilizing differences in energy demand between jobs (depending on machine and product). Jobs are deferred or assigned to another machine. The target energy schedule exhibits an hourly resolution as hourly varying electricity prices are used to reduce electricity costs.

Based on detailed review of existing approaches and against the objective to develop a novel concept for energy flexibility of manufacturing systems, evaluation criteria are defined within the next section to formally evaluate relevant existing research and identify prevalent research gaps and thus research demand.

3.5 Evaluation of Approaches

After providing necessary background detail on existing relevant approaches and discussing their methodology, a formal evaluation scheme is defined. This scheme is then applied to discussed methods and based on the evaluation outcome, research gaps, and thus existing research demand is identified.

3.5.1 *Evaluation Criteria*

Evaluation criteria are defined according to current state of technology and research. The theoretical (technical) background provides an indication on what (broader) dimensions and topics need to be addressed and what elements should be included to establish a sound structure. Findings from existing and reviewed research are then used to further detail evaluation dimensions and sub-dimensions and to define fulfillment levels for evaluation.

First, criteria are defined and their brief description included before evaluating existing approaches according to defined criteria. The research review as well as existing technology revealed that a multitude of different methodological approaches, combining several scientific and technological disciplines, exist. Therefore, the following criteria aim at including all relevant aspects for developing a new energy flexibility concept for manufacturing systems, and a granular and detailed criteria structure has been chosen to account for the variety of existing methods. Three main categories are defined: **I General**, **II System and structure** and **III Advancement**.

The following elements are used to evaluate reviewed approaches within the first category, **I General**, which summarizes general aspects such as overall energy flexibility relation and an approach's planning horizon and temporal dynamics:

I.a General

I.a.i Assessment and improvement: Evaluates if a qualitative or quantitative assessment of energy flexibility takes place or if the method actively improves energy flexibility, potentially enacting multiple levers.

I.a.ii Modularity: To what degree the approach is based on a modular structure which allows flexible enabling and evaluation of components, i.e., including and excluding (manufacturing) system or method's elements.

I.b Planning horizon

I.b.i Reflection of managerial planning horizons: If different managerial planning horizons are included, from tactical to execution, and if dependencies between different time horizons are reflected.

I.b.ii Degree of tactical planning: To what extent tactical planning tasks are reflected and used for energy flexibility purposes.

I.b.iii Degree of operational planning: Equivalent to *I.b.ii Degree of tactical planning*.
I.b.iv Degree of execution: Equivalent to *I.b.ii Degree of tactical planning*.

I.c Temporal dynamics

I.c.i Changeability of (operational) planning horizon: Describes if the option of variable planning horizon definition is available, i.e., one shift versus one year.
I.c.ii Detail of time resolution: Evaluates how granular the proposed time resolution is (in the base case).
I.c.iii Variability of temporal resolution: Evaluates if the temporal resolution is changeable, i.e., from hours to minutes and vice versa.
I.c.iv Influence of initial and final model state: Evaluation and removal of dependency from initial and final state, i.e., buffer inventories or beginning machine states.

Within the category **II System and structure**, considered electricity supply and system elements, including supply terms (if any) and temporal resolution of electricity supply time series are included. Further, different characteristics of a manufacturing system are evaluated, including dynamic interaction, auxiliary system, the systems structure and complexity as well as detail on processes (e.g., different type and electricity demand dynamics):

II.a Electricity system

II.a.i Changeability of supply parameters and structure: Indicates if the approach allows application of different electricity supply scenarios, i.e., multiple DR events, different contract terms, multiple supply sources, etc.
II.a.ii (Contract) supply terms: If and how supply terms are reflected, including, for example, varying prices, maximum contracted supply, maximum demand charges.
II.a.iii Detail of supply temporal resolution: Describes how granular (time) electricity supply is modeled, i.e., if a time series with high resolution (minutes or seconds) can be applied or if only a less detailed time series can be included (e.g., hourly).
II.a.iv Variability of supply temporal resolution: Considers if the applied temporal resolution can be changed, i.e., if the approach/method is capable to incorporate minute and hourly data, or if the resolution is fixed.
II.a.v Decentralized small-scale energy storage: Decentralized small-scale energy storage (e.g., batteries, capacitors, flywheels) are regarded a promising method to integrate VRE, also in combination with other measures (compare Sect. 2.3.1). Therefore, this dimension evaluates if decentralized small-scale energy storage is included and to what degree temporal dynamics and interaction with other system elements are considered.

II.b Manufacturing system

II.b.i Manufacturing system interaction: Describes the level of detail for the modeled manufacturing system and especially interaction of components: number of processes (one- or multi-process system) and if dependencies between processes are considered (e.g., limited buffer capacity between processes), adequate reflection of inter-temporal dynamics and dependencies.
II.b.ii Auxiliary system interaction: To which degree the manufacturing system interacts with the auxiliary system, to which degree the auxiliary system itself is part of proposed energy flexibility method and the level of included detail (parameters, behavior) for the auxiliary system.
II.b.iii Process structure complexity: If and to what complexity process structures can be reflected, e.g., multiple processes, diverting and converging flows, parallel flows.
II.b.iv Variable process types: Reflection of different process types, e.g., continuous and binary (produce/idle) processes and configuration options, e.g., basic (on/off) configuration or advanced configuration (e.g., flexible electricity demand).
II.b.v Electricity demand dynamics: To which degree electricity demand dynamics are reflected, i.e., state-dependent demand or production rate-dependent demand, possibility to include multiple influencing factors on electricity demand (product-dependent, manufacturing program dependent).

For the third category, **III Advancement**, different evaluation dimensions which are included in approaches are characterized, e.g., operational, economic, and environmental evaluation. Further, evaluation of (energy flexibility) improvement methods and their detail is evaluated, as well as applicability of proposed methodology:

III.a Evaluation

III.a.i Operational evaluation: If and how operational factors are considered within evaluation of an approach's effectiveness and impact. Detail of evaluation, i.e., qualitative discussion or quantitative estimate, and if all relevant influencing factors are considered and results provide a sound basis for decision-making.
III.a.ii Economic evaluation: Equivalent to *III.a.i Operational evaluation*.
III.a.iii Environmental evaluation: Equivalent to *III.a.i Operational evaluation*.

III.b Improvement

III.b.i Adjustability of input parameters and structure: Amount of different (adjustable) input parameters and if and to which degree overall system structure is adjustable.
III.b.ii Flexibility of objective definition: If and how many objectives (e.g., minimize cost and energy) can be defined and evaluated and if objective(s) can be adjusted and/or defined as needed.

III.b.iii Input uncertainty and sensitivities: Consideration of uncertainty in input variables (parameters): deterministic evaluation, sensitivity analysis, or consideration of probabilistic variables.

III.c Applicability

III.c.i Visualization of results: If and how obtained results are communicated and visualized, from simple data representation in numbers to dynamic chart visualization.
III.c.ii Implemented solution: Availability of an implemented solution: implemented into (software) prototype which is easy, safe, and fast to adapt for an outside user.
III.c.iii Easiness of configuration: If model or prototype can intuitively configured including fast, safe, and easy changes of elements and structure.
III.c.iv Easiness of parameterization: Speed and required effort for system and scenario parameterization.
III.c.v Intuitively and safe to use: Describes how easy the approach and/or proposed methodology can be applied by an (external) user. Includes required effort to understand and apply the method, and if the method is safe to use (i.e., no detailed knowledge about assumptions required and/or integrity checks automatically performed).

Each criterion dimension is qualitatively evaluated using Harvey Balls from empty to full in quarterly steps (○, ◔, ◑, ◕, ●). Tables A.1, A.2 and A.3 in Sect. A from the appendix summarize scoring examples for each evaluation criterion and scoring value to facilitate and increase objectivity of research evaluation. A higher score (added quarter to a Harvey Ball) implies that all criteria for lower scoring are also fulfilled (cumulative scoring, cf. examples within tables). The given examples are included to provide guidance for scoring. Nonetheless, being a qualitative approach and considering the diversity of reviewed research, not all criterion characteristics can be unambiguously found in reviewed approaches and scoring has been made with regards to best knowledge and belief to allow for a sound evaluation and comparison of results.

3.5.2 *Evaluation of Approaches*

Detailed evaluation of all approaches discussed in Sect. 3.4, based on defined criteria, is summarized in Table 3.3.[3] The overall percentage scoring (100 % is equivalent to a full Harvey Ball for all approaches) aims at providing an indication which criterion is fulfilled to a greater extent than another criterion to identify research demand and visualize differences, but does not claim to allow derivation of a certain

[3]Note that additions (especially a GUI and implementation through PLCs) to Babu and Ashok (2008) from Ashok (2010) and Mohan and Ashok (2011) have been reflected in the evaluation of Babu and Ashok (2008).

quantitative comparison (i.e., "criterion x is 20 % more fulfilled than criterion y" is not a suitable interpretation, but "criterion x is, on average, more often or more thoroughly addressed than criterion y"). Moreover, the stand-alone percentage score for a criterion illustrates how far away existing research is, on average, to fully reflect respective criterion. As mentioned before, criteria and scoring were designed for a cumulative evaluation, i.e., reaching a higher score implicitly assumes that the criteria for the lower score were also met.

3.6 Derivation of Research Demand

After the comprehensive discussion and evaluation of existing research approaches, this section provides a structured summary of prevailing research demand. Before discussing detailed research demand, the following general conclusions can be drawn:

- Table 3.3 shows selective blank spaces, both within major categories and also within single criterion dimensions. However, most criteria are (at least partly) addressed by multiple approaches. As criteria have been defined based on the review of existing research within a narrow field (energy flexibility of manufacturing systems), all dimensions should be addressed by available approaches (which reflects a sound definition of evaluation criteria). Nonetheless, most criteria are not sufficiently/not in detail addressed (several quarter and half-full Harvey Balls, cf. percentage values in the last column). Further, no single approach exists which covers all or most of the criteria to a satisfying extent. In summary, no approach exists which fully addresses relevant criteria, while no criterion is fully addressed by several/most approaches, leaving the demand for an approach which covers all/most criteria in combination.
- Approaches are classified with respect to their (main) planning horizon and process structure. As such, different categories need to be also compared, e.g., several multi-stage operational planning approaches include lots of detail on electricity systems, while execution-related multi-stage approaches include far less detail. Consequently, adding detail for the electricity system is not a general research gap, but specific if an execution-related approach is developed.

The authors in Giret et al. (2015) provide an additional indication towards research demand: they note in their comprehensive review on manufacturing scheduling towards sustainability that most existing approaches focus on offline determination of manufacturing schedules and neglect required reactions (i.e., utilizing online information). Further, considering multiple planning horizons, and subsequent integration into existing planning and scheduling infrastructure is also highlighted for future research. This includes the development of reactive planning approaches (i.e., online) and especially hybrid (online and offline) planing approaches.

Table 3.3 Evaluation of existing research approaches related to energy flexibility of manufacturing systems

Classi-fication	Cate-gory	Criterion	Graßl (2015)	Daryanian (1989)	Yusta (2010)	Wang (2011)	Sharma (2015)	Ashok (2000)	Ashok (2001)	Ashok (2006)	Babu (2008)	Middelberg (2009)	Emec (2013)	Fernandez (2013)	Wang (2013)	Bego (2014)	Sun (2015)	Keller (2015)	Popp (2014)	Li (2012a)	Zhou (2013)	Sun (2014)	Schultz (2015)	Average
		Planning horizon	Tactical	Operational				Operational											Exec.	Execution				
		Process system type	Single	Single				Multi											Single	Multi				
I General	I.a General	I.a.i Assessment and improvement	●	◑	◑	◑	◑	◕	●	◕	◕	◑	◕	◕	●	◕	●	◕	◔	◑	◕	◕	◕	70%
		I.a.ii Modularity	●	○	○	○	◔	○	◑	◔	○	○	◔	○	○	○	◑	○	○	◑	○	○	○	16%
	I.b Planning horizon	I.b.i Reflection of managerial planning horizons	○	◑	◔	◔	◔	◔	◑	◑	◔	◔	◔	◑	◔	◔	◔	◕	◔	◔	◔	◔	◑	31%
		I.b.ii Degree of tactical planning	◑	◔	○	○	○	○	◔	◔	○	○	○	◔	○	◔	○	◔	○	○	○	○	○	10%
		I.b.iii Degree of operational planning	○	◕	◕	◔	◑	◔	◕	◕	◑	◑	◑	◕	◑	◑	◑	◕	◔	○	○	○	◔	44%
		I.b.iv Degree of execution	○	○	○	◔	◔	○	○	○	○	○	○	○	○	○	○	◔	◑	◑	◑	◑	◑	14%
	I.c Temporal dynamics	I.c.i Changeability of (operational) planning horizon	◔	◑	◔	◑	◑	◔	◔	◔	◔	◕	◑	◔	◑	◕	◔	◑	◔	◔	◑	◔	○	39%
		I.c.ii Detail of time resolution	●	◔	◔	◕	◑	◔	◔	◑	◔	◔	◑	●	◔	◔	◔	◔	●	●	●	●	◔	54%
		I.c.iii Variability of temporal resolution	○	◔	◔	◔	◔	◔	◑	◑	◕	◑	◔	◕	◔	◑	◔	◔	●	◕	●	◑	◔	45%
		I.c.iv Influence of initial and final model state	○	●	○	○	◔	○	○	○	○	○	◑	◕	◕	○	○	○	○	○	○	○	○	16%
II System and structure	II.a Electricity system	II.a.i Changeability of supply parameters and structure	○	◕	◕	◑	◔	◑	◑	◑	◑	◑	◑	◑	◑	◕	◔	◑	○	◔	◑	◑	◔	45%
		II.a.ii (Contract) supply terms	○	◔	◔	○	◑	◔	◑	◑	◕	◕	◔	◑	◑	◕	◑	◑	○	○	○	○	○	34%
		II.a.iii Detail of supply temporal resolution	○	◑	◑	◑	◑	◑	◑	◑	◔	◑	◑	●	◑	◑	◑	◑	○	○	◕	◕	◑	46%
		II.a.iv Variability of supply temporal resolution	○	◔	◔	◑	◑	◔	◔	◔	◑	◑	◑	◔	◔	◑	◑	◔	○	◑	◕	◑	◔	36%
		II.a.v Decentralized small-scale energy storage	○	○	○	◑	○	○	○	○	○	○	○	○	○	○	○	◔	○	○	○	○	◔	4%
	II.b Manufacturing system	II.b.i Manufacturing system interaction	◔	○	○	○	◑	◑	◑	◕	◕	◑	◑	◕	◑	◑	◑	◕	◔	◕	◕	◕	◕	49%
		II.b.ii Auxiliary system interaction	○	○	○	○	○	◔	◔	◔	○	○	○	○	○	○	◕	○	●	○	○	○	○	13%
		II.b.iii Process structure complexity	○	○	○	○	◔	◔	◑	◑	●	◔	◔	◔	◔	◔	◔	◑	○	◑	◑	◔	◑	29%
		II.b.iv Variable process types	●	○	◔	○	◑	○	○	◕	◔	○	○	◔	○	○	○	◔	●	◔	○	◔	◔	24%
		II.b.v Electricity demand dynamics	◕	◑	◕	◑	◕	◔	◔	◕	◔	◔	◕	◔	◔	◔	◔	◑	◕	◔	◔	◕	◕	48%
III Advancement	III.a Evaluation	III.a.i Operational evaluation	●	◔	◔	◔	◑	◔	◔	◔	◑	◑	○	◑	◕	◑	◑	○	◔	◑	◑	◑	◑	40%
		III.a.ii Economic evaluation	◕	◔	◑	○	◑	◑	◑	◑	◑	◑	◑	◕	◕	◕	◑	◕	○	○	◔	○	○	44%
		III.a.iii Environmental evaluation	○	○	○	◕	◑	○	○	○	○	○	○	○	○	○	○	○	○	○	○	○	○	6%
	III.b Improvement	III.b.i Adjustability of input parameters and structure	○	◔	◔	◔	◑	◔	◑	◑	◕	○	◔	◔	◕	◕	◔	◕	◔	◕	◔	○	◔	30%
		III.b.ii Flexibility of objective definition	○	◔	◔	◔	◕	◔	◔	◔	◔	◔	◔	◔	◑	◔	◔	◔	◔	◕	◔	◔	◔	30%
		III.b.iii Input uncertainty and sensitivities	○	◑	◑	○	◔	○	○	◔	◑	○	◔	◕	◑	◔	○	◔	◔	◕	◑	◔	◕	29%
	III.c Applicability	III.c.i Visualization of results	◑	◑	◑	◔	◑	◑	◑	◑	◑	◑	◑	◔	◕	◑	◑	◑	◑	◑	◕	◑	◑	50%
		III.c.ii Implemented solution	○	◑	◑	◑	◑	◔	◔	◔	◔	◔	◑	◑	◑	◔	◔	◑	◔	◔	◔	◔	◑	34%
		III.c.iii Easiness of configuration	◑	◔	◔	◔	◔	◔	◔	◔	◕	◔	◑	◔	◔	◔	◔	◔	○	◔	◕	◔	◔	31%
		III.c.iv Easiness of parameterization	◕	◑	◑	◔	◔	◑	◔	◔	◕	◑	◑	◔	◑	◑	◑	◔	◑	◔	◕	◔	◔	44%
		III.c.v Intuitively and safe to use	◑	◑	◔	◔	◔	◔	◔	◔	◕	◔	◔	◔	◔	◑	◑	◔	◑	◔	◔	◔	◔	34%

Taking all previously discussed findings into account, the following detailed research demand can be identified:

D1 A **quantitative approach** for evaluation and methods to enable energy flexibility is required. Even though most approaches are of quantitative nature, some approaches exist which lack systematic criteria for quantitative evaluation and improvement.

D2 A **modular approach** is required to manage complexity and address needs of multiple different applications. A modular structure allows implementation and testing of separated modules, which generally requires less resources compared to an integrated testing and implementation of a complex approach or model. Further, with regards to applicability, especially existing research focuses on selected processes, industries or electricity supply options. In order to be applicable for a wide range of manufacturing systems and electricity supply scenarios, a modular approach would allow enabling, adapting and configuring modules which are tailored to a given application.

D3 Different and **integrated managerial planning horizons** need to be reflected. While methods and tools for each planning horizon are required to fulfill specific tasks, their interaction and mutual influences need consideration to reflect reality and fully exploit the potential for energy flexibility improvement. Further, aside from mutual dependencies, each single planning stage is required in a state-of-the-art management system.

D4 A long-term and changeable **planning horizon** (different duration and granularity, depending on planning task) is required due to (a) the intermittent characteristic of VRE, which can include secondly, hourly, daily and seasonal fluctuations (see also Sect. 2.2.3), (b) operation of manufacturing systems can underlie short- and medium-term influences, e.g., shift times, non-weekend work and stochastic failures, (c) long-term quantification to adequately quantify stochastic behavior and reduce influences of initial and final state.

D5 A changeable and granular **temporal resolution** ensures that both manufacturing system and volatile energy supply dynamics are adequately reflected. For example, existing approaches with a temporal resolution of one hour inherently neglect minute variations of VRE combined with multi-second process cycle times. Consequently, temporal resolution needs to sufficiently reflect physical supply characteristics.

D6 Full **system dynamics**, including interaction of all elements and reflection of initial and final state (discrete and continuous elements) need to be represented. Especially discrete manufacturing systems combined with continuous energy flows require a mixed approach. Further, dynamic dependencies (manufacturing system and energy supply) need to be appropriately considered.

D7 Including an **extended DR spectrum**, i.e., for more than a single event or change in electricity supply parameters (TOU rates or CPP event) is needed. Being able to consider a variety of DR options (price- and event-driven, including RTP) in combination with on-site decentralized generation would significantly increase any approach's applicability and contribution towards energy flexibility.

D8 Uncertainty and **variability of energy supply** should be adequately reflected, including possible strong and immediate fluctuations and unpredictability/ forecasting errors of VRE supply.

D9 Including dynamic interaction with a **decentralized, small-scale energy storage** option accounts for current development towards (decentralized) deployment of small-scale energy storage. Flexible inclusion would allow estimating the impact and benefit of decentralized energy storage for VRE integration. Existing research largely neglects inclusion of energy storage options.

D10 Consideration of a (complex) **manufacturing system** and not limiting evaluation and optimization to separated processes or single machines is required to make an approach being applicable for most (discrete) manufacturing cases (interlinked manufacturing lines and systems with different product variants, etc., see also 2.1.4).

D11 Integrated consideration of (all) **material and energy flows** to not neglect dynamics, influences and avoid isolated optimization of a single (material or energy) flow.

D12 An adequate **process structure complexity** should be representable. Existing approaches, for example, optimization approaches, consider manufacturing lines with a limited number of processes and diverting/ combining flows. Depending on manufactured product and utilized production technology, more complex process structures need to be considered to enable an integrated approach which is widely applicable.

D13 **Auxiliary systems** are rarely considered in existing research, despite their potentially significant contribution to total energy demand of factories and their interaction with manufacturing machines and systems and potential for enabling further energy flexibility (see, e.g., Beier et al. 2015 for an example using flexible CA generation for energy flexibility improvement).

D14 Especially for optimization problems (i.e., NIP, MILP, MINLP), **computational complexity** hinders applicability and implementation. If time and required resources exceed a given threshold, a method becomes inapplicable for a given (real-life) problem. For the case of discrete time formulations, increasing timely resolution (i.e., minutes instead of hours) or the targeted planning horizon (i.e., from one day to one week) can significantly increase computational complexity. As a result, a new approach should allow an efficient solution (within reasonable time and resources) to real-life problems and their complexity.

D15 Available **execution** approaches are centered around classical (emergency) DR events and mostly include a single event within their planning horizon. Combined with the availability of multiple planning and optimization methods, execution approaches using online information lack extensive scientific investigation. In addition, already mentioned uncertainty of VRE generation, forecasting errors and computational complexity combined with complex system structures and long-term planning horizons indicate that execution approaches are required for future integration of VRE.

D16 Integrated evaluation of **multiple indicators** such as operational KPIs, economic factors and ecological impact is required to enable multi-criteria decision-making. Consideration of multiple criteria is rarely performed but required to evaluate if and to what extent improvement measures induce target conflicts (e.g., improved ecological impact but higher cost).

D17 Full **economic evaluation** including all relevant (electricity) supply contract terms, inventory holding cost, capital investment or throughput loss is not performed in any existing approach. Nonetheless, an economic impact evaluation is required, especially when sacrificing throughput for load shifting purposes, also to enable implementation (Paulus and Borggrefe 2011; Sun and Li 2014).

D18 **Visualization of results** is mostly realized by reporting of results in tables or charts. Only few examples exist where an interactive and/or dynamic and comprehensive visualization of results is implemented.

D19 Straightforward **configuration and parameterization** of approaches and methods is a requirement for widespread implementation. Nonetheless, most approaches lack any description of implemented solutions and their configurability.

D20 (Automated) and uncomplicated evaluation of different **scenarios** is the foundation for evaluating improvement measures. Existing approaches have a limited number of input parameter sets and system structure changes. Further, required effort (resources and time) for scenario evaluation and comparison is regularly not discussed in detail.

D21 Existing approaches which focus on energy flexibility mostly investigate a single option to enable or improve energy flexibility. Nonetheless, a combined consideration of **multiple levers** should be performed to avoid neglecting mutual (positive and negative) dependencies.

D22 **Input uncertainty and sensitivities** are, for most existing methods, neither discussed nor quantified. Evaluation of input uncertainties and stochastic influences on nonlinear dynamic models provides additional insight into model behavior and increases proximity to reflect real behavior of components.

In summary, 22 different research demand items to direct future research have been identified. Some of the identified topics depend on each other, for example, a long planning horizon influences computational complexity. Consequently, a novel approach needs to ensure that all items are jointly included to ensure that identified research demand is addressed. The following chapter outlines a new concept and detailed steps and methods for energy flexibility control and improvement of manufacturing systems.

References

Abele, E., Braun, S., & Schraml, P. (2015a). Holistic simulation environment for energy consumption prediction of machine tools. *Procedia CIRP*, *29*, 251–256.

Abele, E., Panten, N., & Menz, B. (2015b). Data collection for energy monitoring purposes and energy control of production machines. *Procedia CIRP*, *29*, 299–304. doi:10.1016/j.procir.2015.01.035.

Alcázar-Ortega, M., Álvarez-Bel, C., Domijan, A., & Escrivá-Escrivá, G. (2012). Economic and environmental evaluation of customers' flexibility participating in operation markets: Application to the meat industry. *Energy*, *41*(1), 368–379. doi:10.1016/j.energy.2012.03.003.

Alcázar-Ortega, M., Álvarez-Bel, C., Escrivá-Escrivá, G., & Domijan, A. (2012b). Evaluation and assessment of demand response potential applied to the meat industry. *Applied Energy*, *92*, 84–91.

Arteconi, A., Hewitt, N., & Polonara, F. (2012). State of the art of thermal storage for demand-side management. *Applied Energy*, *93*, 371–389.

Ashok, S. (2006). Peak-load management in steel plants. *Applied Energy*, *83*(5), 413–424.

Ashok, S. (2010). PLC-based load management in steel rolling mills. *Energy Engineering*, *107*(4), 38–53.

Ashok, S., & Banerjee, R. (2000). Load-management applications for the industrial sector. *Applied Energy*, *66*(2), 105–111.

Ashok, S., & Banerjee, R. (2001). An optimization mode for industrial load management. *IEEE Transactions on Power Systems*, *16*(4), 879–884. http://ieeexplore.ieee.org/lpdocs/epic03/wrapper.htm?arnumber=962440.

Ashok, S., & Banerjee, R. (2003b). Optimal operation of industrial cogeneration for load management. *IEEE Transactions on Power Systems*, *18*(2), 931–937.

Ashok, S., & Banerjee, R. (2003a). Optimal cool storage capacity for load management. *Energy*, *28*(2), 115–126.

Babu, C., & Ashok, S. (2008). Peak load management in electrolytic process industries. *IEEE Transactions on Power Systems*, *23*(2), 399–405. http://ieeexplore.ieee.org/lpdocs/epic03/wrapper.htm?arnumber=4484953.

Babu, C., & Ashok, S. (2009). Optimal utilization of renewable energy-based IPPs for industrial load management. *Renew Energy*, *34*(11), 2455–2460. http://linkinghub.elsevier.com/retrieve/pii/S0960148109001013.

Balogun, V. A., & Mativenga, P. T. (2013). Modelling of direct energy requirements in mechanical machining processes. *Journal of Cleaner Production*, *41*, 179–186. http://linkinghub.elsevier.com/retrieve/pii/S0959652612005471.

Bego, A., Li, L., & Sun, Z. (2014). Identification of reservation capacity in critical peak pricing electricity demand response program for sustainable manufacturing systems. *International Journal of Energy Research*, *38*(6), 728–736. http://onlinelibrary.wiley.com/doi/10.1002/er.3077/full.

Beier, J., Thiede, S., & Herrmann, C. (2015). Increasing energy flexibility of manufacturing systems through flexible compressed air generation. *Procedia CIRP*, *37*, 18–23.

Beier, J., Thiede, S., & Herrmann, C. (2017). Energy flexibility of manufacturing systems for variable renewable energy supply integration: real-time control concept and simulation. *Journal of Cleaner Production*, *141*, 648–661.

Bougain, S., Gerhard, D., Nigischer, C., & Ugurlu, S. (2015). Towards energy management in production planning software based on energy consumption as a planning resource. *Procedia CIRP*, *26*, 139–144. http://www.sciencedirect.com/science/article/pii/S2212827114009068.

Braun, J. E. (2003). Load control using building thermal mass. *Journal of Solar Energy Engineering*, *125*(3), 292–301.

Brundage, M. P., Chang, Q., Li, Y., Arinez, J., & Xiao, G. (2014). Utilizing energy opportunity windows and energy profit bottlenecks to reduce energy consumption per part for a serial production line. In *2014 IEEE International Conference on Automation Science and Engineering* (pp. 461–466), Taipei.

Bruzzone, A. A. G., Anghinolfi, D., Paolucci, M., & Tonelli, F. (2012). Energy-aware scheduling for improving manufacturing process sustainability: A mathematical model for flexible flow shops. *CIRP Annals Manufacturing Technology*, *61*, 459–462. doi:10.1016/j.cirp.2012.03.084.

Castro, P. M., Harjunkoski, I., & Grossmann, I. E. (2009). New continuous-time scheduling formulation for continuous plants under variable electricity cost. *Industrial and Engineering Chemistry Research*, *48*(14), 6701–6714. http://pubs.acs.org/doi/abs/10.1021/ie900073k.

Castro, P. M., Harjunkoski, I., & Grossmann, I. E. (2011). Optimal scheduling of continuous plants with energy constraints. *Computers and Chemical Engineering*, *35*(2), 372–387. http://linkinghub.elsevier.com/retrieve/pii/S0098135410001687.

Central Electricity Authority. (2014). Load Generation Balance Report 2014–15. Technical report, Central Electricity Authority, New Delhi.

Dai, M., Tang, D., Giret, A., Salido, M. A., & Li, W. D. (2013). Energy-efficient scheduling for a flexible flow shop using an improved genetic-simulated annealing algorithm. *Robot Computer Integrated Manufacturing*, *29*(5), 418–429.

Daryanian, B., Bohn, R. E., & Tabors, R. D. (1989). Optimal demand-side response to electricity spot prices for storage-type customers. *IEEE Transactions on Power Systems*, *4*(3), 897–903. http://ieeexplore.ieee.org/lpdocs/epic03/wrapper.htm?arnumber=4310873.

Diaz, N., Ninomiya, K., Noble, J., & Dornfeld, D. (2012). Environmental impact characterization of milling and implications for potential energy savings in industry. *Procedia CIRP*, *1*, 518–523. http://linkinghub.elsevier.com/retrieve/pii/S2212827112000935.

Eberspächer, P., Schraml, P., Schlechtendahl, J., Verl, A., & Abele, E. (2014). A model- and signal-based power consumption monitoring concept for energetic optimization of machine tools. *Procedia CIRP*, *15*, 44–49. doi:10.1016/j.procir.2014.06.020.

Emec, S., Kuschke, M., Huber, F. W., Stephan, R., Strunz, K., & Seliger, G. (2013). Stochastic optimization method to schedule production steps according to volatile energy price. In G. Seliger (Ed.), *GCSM 11th global conference on sustainable manufacturing* (pp. 637–642). Berlin: Springer.

Fang, K., Uhan, N., Zhao, F., & Sutherland, J. W. (2011). A new approach to scheduling in manufacturing for power consumption and carbon footprint reduction. *Journal of Manufacturing Systems*, *30*(4), 234–240. http://linkinghub.elsevier.com/retrieve/pii/S0278612511000690.

Feder, F., Erlach, K., Hosak, F., & Lepple, H. (2015). Der energieflexible und effiziente Energiewertstrom. *wt Werkstattstech online*, *105*(5), 313–318.

Fernandez, M., Li, L., & Sun, Z. (2013). "Just-for-Peak" buffer inventory for peak electricity demand reduction of manufacturing systems. *International Journal of Production Economics*, *146*(1), 178–184.

Finn, P., & Fitzpatrick, C. (2014). Demand side management of industrial electricity consumption: Promoting the use of renewable energy through real-time pricing. *Applied Energy*, *113*, 11–21. doi:10.1016/j.apenergy.2013.07.003.

Finn, P., O'Connell, M., & Fitzpatrick, C. (2013). Demand side management of a domestic dishwasher: Wind energy gains, financial savings and peak-time load reduction. *Applied Energy*, *101*, 678–685. doi:10.1016/j.apenergy.2012.07.004.

FOREnergy (2015). FOREnergy: Die energieflexible Fabrik. Retrieved August 13, 2015, from http://www.forenergy.de/.

Gerhard, D. (2015). Integrated systems: innovations and applications. In M. Fathi (Ed.), *Integrated systems innovations applied* (pp. 29–38). Cham, Switzerland: Springer International Publishing.

Ghadimi, P., Kara, S., & Kornfeld, B. (2015). Renewable energy integration into factories: Real-time control of on-site energy systems. *CIRP Annals Manufacturing Technology*, *64*(1), 443–446.

Ghadimi, P., Li, W., Kara, S., & Herrmann, C. (2014). Integrated material and energy flow analysis towards energy efficient manufacturing. *Procedia CIRP*, *15*, 117–122. doi:10.1016/j.procir.2014.06.010.

Giret, A., Trentesaux, D., & Prabhu, V. (2015). Sustainability in manufacturing operations scheduling: A state of the art review. *Journal of Manufacturing Systems*, *37*(1), 126–140.

Graßl, M. (2015). *Bewertung der Energieflexibilität in der Produktion*. München: Herbert Utz.

Graßl, M., & Reinhart, G. (2014). Evaluating measures for adapting the energy demand of a production system to volatile energy prices. *Procedia CIRP*, *15*, 129–134.

Graßl, M., Vikdahl, E., & Reinhart, G. (2013). A petri-net based approach for evaluating energy flexibility of production machines. In M. Zäh (Ed.), *5th international conference on chang agil reconfigurable virtual production (CARV 2013)* (pp. 303–308). Switzerland, Munich: Springer International Publishing.

Grismajer, M., & Seliger, G. (2012). Information requirements for motivated alignment of manufacturing operations to energy availability. *Procedia CIRP*, *3*(1), 418–423. doi:10.1016/j.procir.2012.07.072.

Gutowski, T. G., Dahmus, J., & Thiriez, A. (2006). Electrical energy requirements for manufacturing processes. In *13th CIRP International Conference on Life Cycle Engineering* (pp. 623–627), Lueven.

Ha, D. L., Joumaa, H., Ploix, S., & Jacomino, M. (2012). An optimal approach for electrical management problem in dwellings. *Energy Build*, *45*, 1–14. doi:10.1016/j.enbuild.2011.11.027.

Heinemann, T., Schraml, P., Thiede, S., Eisele, C., Herrmann, C., & Abele, E. (2014). Hierarchical evaluation of environmental impacts from manufacturing system and machine perspective. *Procedia CIRP*, *15*, 141–146. doi:10.1016/j.procir.2014.06.063.

Jang, D., Eom, J., Kim, M. G., & Rho, J. J. (2015). Demand responses of Korean commercial and industrial businesses to critical peak pricing of electricity. *Journal of Cleaner Production*, *90*, 275–290. http://linkinghub.elsevier.com/retrieve/pii/S0959652614012396.

Junge, M. (2007). *Simulationsgestützte Entwicklung und Optimierung einer energieeffizienten Produktionssteuerung*. Kassel University Press GmbH, Kassel.

Kara, S., & Li, W. (2011). Unit process energy consumption models for material removal processes. *CIRP Annals Manufacturing Technology*, *60*(1), 37–40. http://linkinghub.elsevier.com/retrieve/pii/S0007850611000199.

Keller, F., & Reinhart, G. (2015). Produktionsplanung unter Berücksichtigung des Energiebezugs. *wt Werkstattstech online*, *105*(3), 141–147.

Keller, F., Schönborn, C., & Reinhart, G. (2015). Energy-orientated machine scheduling for hybrid flow shops. *Procedia CIRP*, *29*, 156–161.

Kwag, H. G., & Kim, J. O. (2012). Optimal combined scheduling of generation and demand response with demand resource constraints. *Applied Energy*, *96*, 161–170. doi:10.1016/j.apenergy.2011.12.075.

Langer, T., Schlegel, A., Stoldt, J., & Putz, M. (2014). A model-based approach to energy-saving manufacturing control strategies. *Procedia CIRP*, *15*, 123–128. doi:10.1016/j.procir.2014.06.019.

Li, L., Chang, Q., & Ni, J. (2009). Data driven bottleneck detection of manufacturing systems. *International Journal of Production Research*, *47*(18), 5019–5036.

Li, W., & Kara, S. (2011). An empirical model for predicting energy consumption of manufacturing processes: A case of turning process. *Proceedings of the Institution of Mechanical Engineers Part B, Journal of Engineering Manufacturing*, *225*(9), 1636–1646.

Li, L., Sun, Z., & Tang, Z. (2012a). Real time electricity demand response for sustainable manufacturing systems: challenges and a case study. In *8th IEEE International Conference Automation Science and Engineering* (pp. 353–357), Seoul. http://ieeexplore.ieee.org/lpdocs/epic03/wrapper.htm?arnumber=6386437.

Li, L., Sun, Z., Yang, H., Gu, F., & (2012b). Simulation-based energy efficiency improvement for sustainable manufacturing systems. In *Proceedings of ASME (2012) International Manufacturing Science and Engineering Conference MSEC2012* (pp. 1033–1039). IN: Notre Dame.

Li, W., Winter, M., Kara, S., & Herrmann, C. (2012c). Eco-efficiency of manufacturing processes: A grinding case. *CIRP Annals Manufacturing Technology*, *61*(1), 59–62. http://linkinghub.elsevier.com/retrieve/pii/S0007850612000315.

Li, L., & Sun, Z. (2013). Dynamic energy control for energy efficiency improvement of sustainable manufacturing systems using markov decision process. *IEEE Transactions on Systems, Man, and Cybernetics: Systems*, *43*(5), 1195–1205. http://ieeexplore.ieee.org/lpdocs/epic03/wrapper.htm?arnumber=6519950.

Li, L., Yan, J., & Xing, Z. (2013). Energy requirements evaluation of milling machines based on thermal equilibrium and empirical modelling. *Journal of Cleaner Production*, *52*, 113–121. http://linkinghub.elsevier.com/retrieve/pii/S0959652613001418.

Liebl, C., & Zäh, M. F. (2014). Energiebedarfssimulation von Produktionsanlagen. *wt Werkstattstech online*, *104*(5), 325–329.

Liebl, C., Popp, R., & Zäh, M. F. (2015). Energieflexibilität von Produktionsanlagen. *wt Werkstattstech online*, *105*(3), 136–140.

Liu, X., Zou, F., & Zhang, X. (2008). Mathematical model and genetic optimization for hybrid flow shop scheduling problem based on energy consumption. In *2008 Chinese Control and Decision Conference* (pp. 1002–1007). http://ieeexplore.ieee.org/lpdocs/epic03/wrapper.htm?arnumber=4597463.

Liu, N., Zhang, Y., & Lu, W. (2015). A hybrid approach to energy consumption modelling based on cutting power: a milling case. *Journal of Cleaner Production*, *104*, 264–272. http://linkinghub.elsevier.com/retrieve/pii/S095965261500596X.

Luo, H., Du, B., Huang, G. Q., Chen, H., & Li, X. (2013). Hybrid flow shop scheduling considering machine electricity consumption cost. *International Journal of Production Economics*, *146*(2), 423–439. doi:10.1016/j.ijpe.2013.01.028.

Middelberg, A., Zhang, J., & Xia, X. (2009). An optimal control model for load shifting - With application in the energy management of a colliery. *Applied Energy*, *86*, 1266–1273.

Mitra, S., Grossmann, I. E., Pinto, J. M., & Arora, N. (2012). Optimal production planning under time-sensitive electricity prices for continuous power-intensive processes. *Computers and Chemical Engineering*, *38*, 171–184. http://linkinghub.elsevier.com/retrieve/pii/S0098135411003012.

Mitra, S., Sun, L., & Grossmann, I. E. (2013). Optimal scheduling of industrial combined heat and power plants under time-sensitive electricity prices. *Energy*, *54*, 194–211. http://linkinghub.elsevier.com/retrieve/pii/S0360544213001448.

Mohan, L., & Ashok, S. (2011). PLC based peak load control for chemical industries. *World Journal of Engineering*, *8*(1), 35–44.

Neugebauer, R., Putz, M., Schlegel, A., Langer, T., Franz, E., & Lorenz, S. (2012). Energy-sensitive production control in mixed model manufacturing processes. In *19th CIRP International Conference Life Cycle Engineering* (pp. 399–404), Berkeley.

Nilsson, K. (1992). Industrial production planning with optimal electricity cost. *Energy Conversion and Management*, *34*(3), 153–158.

Nilsson, K., & Söderström, M. (1993). Industrial applications of production planning with optimal electricity demand. *Applied Energy*, *46*(2), 181–192.

Outhred, H., Bannister, C., Kaye, R., Lee, Y., Sutanto, D., & Manimaran, R. (1988). Electricity pricing: Optimal operation and investment by industrial consumers. *Energy Policy*, *16*(4), 384–393.

Paulus, M., & Borggrefe, F. (2011). The potential of demand-side management in energy-intensive industries for electricity markets in Germany. *Applied Energy*, *88*(2), 432–441.

Pechmann, A. & Schöler, I. (2011). Optimizing energy costs by intelligent production scheduling. In *Proceedings of the 18th CIRP International Conference of Life Cycle Engineering* (pp. 293–298), Braunschweig.

Popp, R., Keller, F., Atabay, D., Dornmair, R., Buderus, J., & Kohl, J. (2013). Technische Innovationen für die Energieflexible Fabrik. *ZWF Zeitschrift für wirtschaftlichen Fabrikbetr*, *108*(07–08), 556–560.

Popp, R., & Zäh, M. F. (2014). Steuerung des Energiebedarfs von Werkzeugmaschinen. *wt Werkstattstech online*, *104*(6), 413–417.

Posselt, G. (2016). *Towards energy transparent factories*. Cham: Springer International Publishing.

Rager, M. (2008). *Energieorientierte Produktionsplanung: Analyse, Konzeption und Umsetzung* (Gabler ed.). Wiesbaden: Wissenschaft.

Rankin, R., & Rousseau, P. (2008). Demand side management in South Africa at industrial residence water heating systems using in line water heating methodology. *Energy Conversion and Management*, *49*(1), 62–74.

Reinhart, G. & Graßl, M. (2013). Energieflexible Fabriken: Maßnahmen zur Steuerung des Energiebedarfs von Fabriken. In *VDI-Expertenforum Energiemanagement - Wie "smart" managen wir Energie wirklich?*, Karlsruhe. Verein Deutscher Ingenieure -VDI-, Düsseldorf. https://www.vdi.de/fileadmin/vdi_de/news_bilder/News_GEU/Grassl_1_.pdf.

Reinhart, G., Graßl, M., & Datzmann, S. (2014). Methode zur Bewertung der Energieflexibilität. *wt Werkstattstech online*, *104*(5), 313–319.

Reinhart, G., Reinhardt, S., & Graßl, M. (2012). Energieflexible Produktionssysteme. *wt Werkstattstech online*, *102*(9), 622–628.

Reinhart, G., & Schultz, C. (2014). Herausforderungen einer energieorientierten Produktionssteuerung. *ZWF Zeitschrift für wirtschaftlichen Fabrikbetr*, *109*(1–2), 29–33.

Reynders, G., Nuytten, T., & Saelens, D. (2013). Potential of structural thermal mass for demand-side management in dwellings. *Building and Environment*, *64*, 187–199.

Schlechtendahl, J., Eberspächer, P., Schraml, P., Verl, A., & Abele, E. (2016). Multi-level energy demand optimizer system for machine tool controls. *Procedia CIRP*, *41*, 783–788. doi:10.1016/j.procir.2015.12.030.

Schultz, C., Braunreuther, S., & Reinhart, G. (2015a). Integration von Lastmanagementstrategien in die Produktionssteuerung. *ZWF Zeitschrift für wirtschaftlichen Fabrikbetr*, *110*(11), 675–680.

Schultz, C., Keller, F., & Reinhart, G. (2014). Modellierung einer energieorientierten PPS. *wt Werkstattstech online*, *104*(11/12), 771–775.

Schultz, C., Keller, F., Simon, P., Reinhart, G., Dornmair, R., & Wagner, U. (2015b). Wirtschaftlicher Einsatz von Energieflexibilität als Lösungsbaustein der Energiewende. *ZWF Zeitschrift für wirtschaftlichen Fabrikbetr*, *110*(7–8), 416–420.

Schultz, C., Sellmaier, P., & Reinhart, G. (2015c). An approach for energy-oriented production control using energy flexibility. *Procedia CIRP*, *29*, 197–202.

Sharma, A., Zhao, F., & Sutherland, J. W. (2015). Econological scheduling of a manufacturing enterprise operating under a time-of-use electricity tariff. *Journal of Cleaner Production*, *108*, 256–270. http://www.sciencedirect.com/science/article/pii/S095965261500712X.

Shrouf, F., Ordieres-Meré, J., García-Sánchez, A., & Ortega-Mier, M. (2014). Optimizing the production scheduling of a single machine to minimize total energy consumption costs. *Journal of Cleaner Production*, *67*, 197–207. http://linkinghub.elsevier.com/retrieve/pii/S0959652613008780.

Sun, Z., & Li, L. (2013). Opportunity estimation for real-time energy control of sustainable manufacturing systems. *IEEE Transactions on Automatic Science and Engineering*, *10*(1), 38–44. http://ieeexplore.ieee.org/lpdocs/epic03/wrapper.htm?arnumber=6329376.

Sun, Z., & Li, L. (2014). Potential capability estimation for real time electricity demand response of sustainable manufacturing systems using Markov Decision Process. *Journal of Cleaner Production*, *65*, 184–193. http://linkinghub.elsevier.com/retrieve/pii/S0959652613005738.

Sun, Z., Li, L., Bego, A., & Dababneh, F. (2015). Customer-side electricity load management for sustainable manufacturing systems utilizing combined heat and power generation system. *International Journal of Production Economics*, *165*, 112–119. http://linkinghub.elsevier.com/retrieve/pii/S0925527315001176.

Sun, Z., Li, L., Fernandez, M., & Wang, J. (2014). Inventory control for peak electricity demand reduction of manufacturing systems considering the tradeoff between production loss and energy savings. *Journal of Cleaner Production*, *82*, 84–93.

Thiede, S. (2012). *Energy efficiency in manufacturing systems*. Berlin, Heidelberg: Springer.

van Staden, A. J., Zhang, J., & Xia, X. (2011). A model predictive control strategy for load shifting in a water pumping scheme with maximum demand charges. *Applied Energy*, *88*(12), 4785–4794. http://linkinghub.elsevier.com/retrieve/pii/S0306261911004442.

Velchev, S., Kolev, I., Ivanov, K., & Gechevski, S. (2014). Empirical models for specific energy consumption and optimization of cutting parameters for minimizing energy consumption during turning. *Journal of Cleaner Production*, *80*, 139–149. http://linkinghub.elsevier.com/retrieve/pii/S0959652614005903.

Wang, X., Ding, H., Qiu, M., & Dong, J. (2011). A low-carbon production scheduling system considering renewable energy. In *Proceedings of 2011 IEEE International Conference on Service Operations, Logistics and Informatics, SOLI 2011* (pp. 101–106).

Wang, Y., & Li, L. (2013). Time-of-use based electricity demand response for sustainable manufacturing systems. *Energy, 63*, 233–244. http://linkinghub.elsevier.com/retrieve/pii/S036054421300844X.

Weinert, N. (2010). *Vorgehensweise für Planung und Betrieb energieeffizienter Produktionssysteme*. Stuttgart: Fraunhofer.

Weinert, N., Chiotellis, S., & Seliger, G. (2011). Methodology for planning and operating energy-efficient production systems. *CIRP Annals Manufacturing Technology, 60*(1), 41–44. http://linkinghub.elsevier.com/retrieve/pii/S0007850611000163.

Yusta, J., Torres, F., & Khodr, H. (2010). Optimal methodology for a machining process scheduling in spot electricity markets. *Energy Conversion and Management, 51*(12), 2647–2654. http://linkinghub.elsevier.com/retrieve/pii/S0196890410002098.

Zein, A. (2012). *Transition towards energy efficient machine tools*. Berlin, Heidelberg: Springer.

Zhou, Z., & Li, L. (2013). Real time electricity demand response for sustainable manufacturing systems considering throughput bottleneck detection. *IEEE International Conference on Automation Science and Engineering* (pp. 640–644), Madison.

Chapter 4
Manufacturing System Real-Time Energy Flexibility Control and Improvement

Within the first chapter, two research questions have been formulated, the current state of technology has been summarized in Chap. 2, followed by reviewing and summarizing existing research (Chap. 3). As a result, research demand has been identified. Therefore, this chapter starts with a definition of main objectives for a novel concept that aims at fulfilling identified research demand. Concept-specific requirements are derived in a next step. Thereon, a planning and improvement framework to establish the concept's role in a manufacturing company is briefly introduced. The concept's overall structure is provided and its elements and interactions described. The remainder of the chapter is structured along the concept's system behavior model elements, which are qualitatively and mathematically described in each corresponding subsection. The chapter closes with a description of an application cycle and an intermediate concluding summary.

4.1 Concept Objective and Requirements

Within the following, identified research demand is translated into thirteen specific objectives which describe a desired target state and vision of a novel concept. These objectives are then utilized to define a set of 22 concept requirements. Finally, these requirements are the basis (specific tasks) for developing a novel concept to move towards fulfillment of general and concept-specific objectives as well as the formulated research question (Fig. 4.1).

Based on the general objective to increase energy flexibility of manufacturing systems and identified research demand in Sect. 3.6, the following concept-specific objectives are formulated. The concept shall

O1 provide an **integrated strategic framework** for energy flexibility improvement of manufacturing systems.

J. Beier, *Simulation Approach Towards Energy Flexible Manufacturing Systems*, Sustainable Production, Life Cycle Engineering and Management, DOI 10.1007/978-3-319-46639-2_4

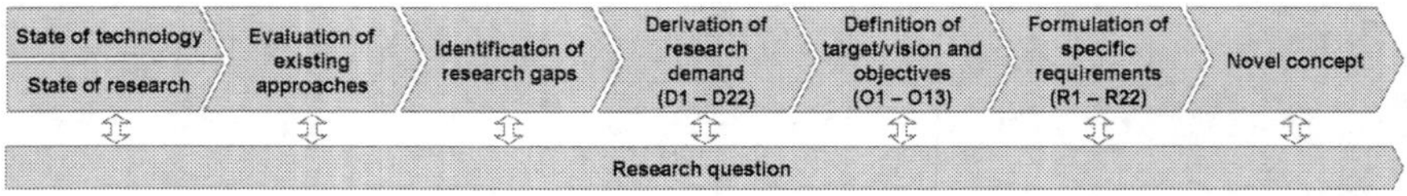

Fig. 4.1 Process for the development of novel concept based on formulated research question, state of technology, and state of research (own illustration)

O2 propose a solution to **control manufacturing systems** under energy flexibility goals.

O3 enable **energy flexibility improvement** through multiple levers (e.g., embodied energy storage, CA storage and batteries) and planning horizons (long-term strategic to mid-/short-term operational and real-time execution) and thus contribute to increasing DR capability of factories.

O4 not be specific to a single or narrow range of task(s) or concept objective(s) but applicable for **multiple, different targets**.

O5 consider all varieties of **temporal dynamics and resolutions** of relevant influencing factors.

O6 be applicable to a wide range of **different manufacturing systems** and consequently products.

O7 be able to consider several **electricity supply options**, including on-site generation, time-dependent availability, and different prices.

O8 include **connected systems** that might have a significant influence on the concept objective or are evaluated for comparison to judge effectiveness of proposed approach, for example, auxiliary systems and direct energy storage options such as batteries.

O9 consider all relevant **material and energy flow dynamics**.

O10 perform a **holistic quantitative evaluation** of multidimensional indicators to provide a sound basis for decision making.

O11 be easy and straightforward **configurable and applicable** for a wide range of applications.

O12 include a **dynamic visualization** of relevant factors to provide a comprehensive overview of input parameters, state-describing, and impact indicators.

O13 introduce **application structure and guidelines**, which are applied to demonstrate applicability of the concept using example application cases.

These objectives describe the desired targets and detailed vision of a new concept to address existing research gaps towards energy flexibility of manufacturing systems. Existing technology and research defines the basis which needs to be extended to reach described desired target state. Consequently, specific requirements are formulated to enable concept development. Requirements are classified into **general**, **electricity supply**, **manufacturing system**, and **improvement and implementation** requirements.

General requirements summarize all overlapping requirements such as integration of the concept into a company's strategy and methodological requirements which emerged from reviewing existing research:

R1 An **energy flexibility vision** as integral part of a company's strategic vision ensures that energy flexibility is treated as a central company goal and is pursuit and supported by decision makers. Without sufficient support throughout relevant hierarchies of a company, implementation is threatened and solutions might not be sustainable. Consequently, the proposed concept should provide means to be integrated into a company's vision and strategy.

R2 A **dynamic system perspective** ensures that no separate improvements are performed, which might neglect mutual influences and/or result in an overall sub-optimal solution. First, temporal dynamics need to be included to adequately reflect characteristics of all concept components (i.e., VRE supply, manufacturing systems), which is not systematically found in existing research. Second, a system perspective is required as selective improvement rarely leads to an overall/system improvement (or lacks the proof thereof). Third, consideration of different planning horizons, from tactical to execution, and their dependencies, ensures that energy flexibility-related planning and control is thoroughly reflected in all managerial planning tasks.

R3 **Modularity** is defined as a general requirement as a modular structure is both required to make the proposed concept universally applicable as well as to ensure efficient validation and easy applicability and configurability to different use-cases. A modular structure should be desired in all concept components. Most existing system approaches exhibit no or do not explicitly mention a modular structural approach.

R4 The requirement of a solution including manufacturing **execution** is mainly derived from existing research demand and stochastic behavior of VRE and manufacturing systems. Aspired concept should provide a solution which can be executed in real-time using online data, and thus provide direct manufacturing system control based on fed back system states. This ensures that stochastic, non-foreseeable behavior can be addressed (i.e., stochastic machine failure or forecasting errors in the context of VRE). Further, only a very limited number of existing approaches consider execution applications.

Based on Sect. 2.2, adequate **electricity supply system** integration and consideration of characteristics result in further requirements. Variability of renewable supply and complexity of existing power systems and supply to industry are reflected in the following requirements:

R5 **Multiple supply sources** should be included within the concept. Developments towards widespread VRE deployment and decentralized electricity generation indicate that future (electrical) energy supply for industry is based on multiple (simultaneously available) sources. Therefore, including relevant supply options is required.

R6 Related to the previous requirement, **complete supply characteristics** such as time-dependent availability and prices, total energy demand, and peak demand limits or overall limited availability are required to sufficiently reflect supply characteristics and evaluate/compare energy flexibility improvement options. In general, relevant parameters include all input factors that are either altered due

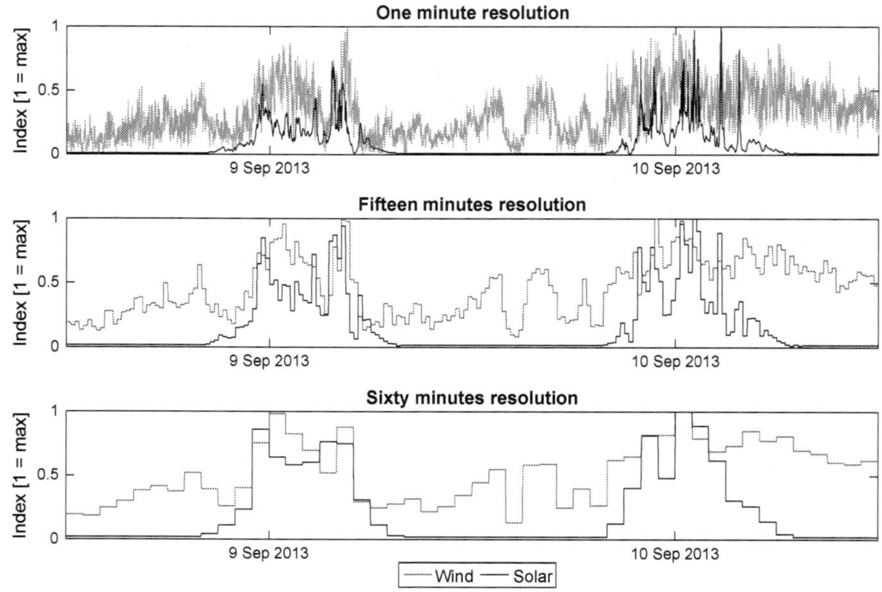

Fig. 4.2 Wind and solar generation at IWF, TU Braunschweig, September 9 and 10, 2013: 1, 15, and 60 min resolution, normalized to sample period's and resolution's maximum generation (own data and illustration)

to decision makers choice (i.e., different supply sources or contracts) or relevant as they are included in comparing energy flexibility measures (i.e., reduced peak demand charges but increased inventory holding cost).

R7 A changeable and sufficiently detailed **temporal resolution** of supply parameters is required to limit estimation errors. Especially, decentralized VRE supply can exhibit significant changes within second intervals. An example is illustrated in Fig. 4.2, which shows wind and solar generation sample data for 2 days with 1, 15, and 60 min resolution. Utilizing averages over longer than minute periods eliminates a large share of inherent supply dynamics, which are in the focus of every energy flexibility approach. Utilizing average values over a given time period implies that either supply was stable or that energy storage means (e.g., capacitors, batteries) exist that have sufficient capacity (power and energy) to smooth variable supply. While this can be assumed to be realistic for small time intervals (e.g., minutes), longer time intervals would require significant energy storage capacity (for examples on wind and solar smoothing see, e.g., Ecoult 2015a, b). In addition, estimating peak demand values correctly, e.g., 15-minute averages, might be challenging if energy and electricity flow resolution is low. Consequently, chosen electricity temporal resolution should sufficiently reflect physical characteristics of VRE and electricity supply in general.

R8 **Decentralized generation and storage** is a requirement resulting from discussed characteristics of VRE supply and integration. Especially for the case of

VRE generation, decentralized solutions are preferable due to available wind and solar sources and their option to directly use power output. Further, to include all relevant supply and demand options for integrating VRE, not only energy flexibility of manufacturing systems, but also decentralized (electrochemical, compressed air) storage and resulting dynamics should be investigated. This includes consideration of EVs as intermittent storage option as EVs might become increasingly available in the (near) future and their utilization dynamics provide additional challenges when used in (decentralized) VRE storage applications.

Manufacturing system requirements are derived from the general understanding of the factory as a dynamic technological system. As discussed in Sect. 2.1, multiple influencing factors such as dynamic component (machines, TBS) interaction and process/machine behavior need to be considered. Consequently, in relation to previously defined objectives and research demand, the following requirements are formulated:

R9 Development and detailed description of a **control algorithm** is required to match electricity demand and variable supply. Manufacturing systems and machines are subject to several technical and operational constraints. Electricity demand of a manufacturing system is dependent on system and individual components' state. Consequently, dynamic electricity demand control under given constraints is an integral part of a new energy flexibility concept.

R10 Dynamic consideration of **manufacturing system interaction** is necessary to estimate material and energy flows. Connected manufacturing systems elements with discrete (i.e., product) and continuous (i.e., electricity) flows directly and indirectly influence other system elements. Evolution of the system state depends on these interactions as well as determining control signals (see previous requirement). As a result, not only single and/or isolated machine or component behavior should be considered but all relevant interactions.

R11 Including relevant **auxiliary system interaction** such as compressed air supply extends the previously discussed requirement of system element interactions. Rationale for inclusion is similar to above manufacturing system interactions, combined with extending the overall system's energy flexibility potential through additional energy storage options.

R12 **Energy and material flows** are essential to determine a system's behavior. All relevant flows across the system boundary (in and out) need to be quantified to allow an evaluation of the system's performance in relation to energy flexibility. To be able to determine these cross-system flows, internal flows need to be dynamically tracked to maintain flow balances. However, insignificant flows (too small or no significant change if system parameters, structure or behavior changes) need to be excluded to limit complexity of the concept.

R13 **Machine behavior**, for example, state-based energy demand and processing/production behavior is the link between energy and material flows as well as the interface for determining and controlling dynamic system evolution and energy demand. As such, an appropriate abstraction level to enable control and flow estimation without compromising easy application and unnecessarily increasing complexity is desired.

R14 Individual **auxiliary system component behavior** and its rationale for inclusion is similar to previous requirement (machine behavior) and emphasizes the role of auxiliary systems in a factory environment.

The central objective of proposed concept is an improvement of energy flexibility, which can be applied and integrated in diverse manufacturing facilities. Based on existing methods and especially open research demand, the following **improvement and implementation** requirements are constituted:

R15 A comprehensive **set of indicators** to allow multidimensional evaluation of the system's state, evolution and performance is needed. This includes operational, economic, and ecological quantitative indicators.

R16 A **cockpit/GUI** enables control and facilitates visualization and thus understanding of system behavior and improvement objectives. Therefore, a structured dynamic reporting of relevant system parameters (e.g., VRE supply), state-describing variables (e.g., current electricity demand) and indicator evolution (e.g., total energy demand) needs to be included in the concept.

R17 **Multi and parallel levers** to improve energy flexibility need consideration to account for dependencies between levers, improvement actions and different planning horizons. First, including only a limited number of improvement options might not exploit the full potential of energy flexibility enhancement. Second, as discussed before, mutual positive and negative influences of actions and planning stages need to be considered. Consequently, various improvement options are to be combined in a new concept.

R18 **Scenario evaluation and comparison** in contrast to evaluating a single set of input parameters provides the basis for systematic improvement. Further, scenario results and output should be standardized and stored in a central database to allow easy evaluation, comparison, and interpretation.

R19 A **structured target search** towards energy flexibility improvement should be available to systematically determine influencing factors. Hypotheses for energy flexibility improvement can then be formulated and tested using multiple sets of input factors (and/or system structures), results evaluated and, dependent on outcome, a new series of input factors can be defined.

R20 **Straightforward, fast, and save set up** of a model based on proposed logical concept is a key factor for application and improvement. This includes providing a structured/modular approach and an explicit description of the concept's elements. A working prototype that provides an implementation example for future use as well as mechanisms to ensure logical integrity (for example, self-validating logical structures) contribute to this requirement.

R21 Following the previous requirement, **fast and easy parameterization** of a ready, structurally set-up model is required to save resources and to enable multi-scenario evaluation and target search. In addition, implementation is facilitated.

R22 A key factor for application and implementation of a new quantitative concept is limiting **computational complexity** to a degree where executing or solving a given real-life problem is feasible within reasonable time and resources. As the concept aims at proving both a real-time applicability, system control approach and energy flexibility improvement, computational complexity depends on the specific use-case and objective. However, two guiding principles are: (1) the problem needs to be solvable, i.e., an optimization problem with exponentially increasing complexity might be considered non-solvable for a (proven) optimization for very large problems due to time and resource requirements and (2) resources (methods, software, hardware) for implementing the quantitative method should be in line with commonly available resources in industry to allow fast adaption and implementation.

Table 4.1 shows the relation of concept-specific objectives and requirements. As discussed before, both objectives and requirements are strongly interrelated and exhibit mutual dependencies (compare also to Fig. 4.1). As such, the presented table aims at providing a qualitative summary, i.e., some objectives might also influence requirements which are not indicated in the table as the influence is relatively small compared to other requirements/objectives. In summary, all discussed concept-specific objectives are reflected in requirements and all requirements are a result of a concept-specific objective. The amount of dots/circles should not be used to derive a priority of objectives or requirements as the assessment is purely qualitative and includes no indication which requirements are a prerequisite for another requirement. The next section presents the detailed methodological approach towards energy flexibility improvement to address identified requirements.

4.2 Concept Overview

Proposed concept structure is based on the production management and control theory and holistic factory understanding outlined in Sect. 2.1 2 and integrates all previously described requirements into a logical framework. Figure 4.3 illustrates the combination of the different production management planning horizons and the holistic factory understanding. In essence, the production management (black box) from the holistic factory understanding (Fig. 2.3) is detailed and expanded by including the different management system elements from Fig. 2.4. The manufacturing management[1] control cycle (control loop one) is explicitly considered, which aims at controlling and measuring feedback from the performance system, and which reflects the different managerial planning horizons to influence the performance system. In turn, the performance system is detailed and includes elements from the holistic factory understanding, namely the manufacturing system, technical building services, and energy and manufacturing control (execution) as main elements.

[1]The term *manufacturing management* is used here instead of the more general term *production management* introduced in Chap. 2 (theoretical background) to highlight the manufacturing focus.

Table 4.1 Relation of concept-specific objectives and resulting requirements. A full dot indicates a direct/strong relation or dependence, an empty dot a loose relation/weak influence

	R1 Energy flexibility vision	R2 Dynamic system perspective	R3 Modularity	R4 Execution	R5 Multiple supply sources	R6 Complete supply characteristics	R7 Temporal resolution	R8 Decentralized generation and storage	R9 Control algorithm	R10 Manufacturing system interaction	R11 Auxiliary system interaction	R12 Energy and material flows	R13 Machine behavior	R14 Auxiliary system component behavior	R15 Set of indicators	R16 Cockpit/GUI	R17 Multi and parallel levers	R18 Scenario evaluation and comparison	R19 Structured target search	R20 Straightforward, fast and save set up	R21 Fast and easy parameterization	R22 Computational complexity
O1 Integrated strategic framework	●	○		○											○					○		
O2 Control manufacturing systems		○	○	●					●	○	○		●	○		○				○		○
O3 Energy flexibility improvement	●	○		○	○				●						○		●	●	●	○	●	●
O4 Multiple, different targets			●		●	●									○		●	○	○	○	○	
O5 Temporal dynamics and resolution		●		●			●			○	○	●	○	○								●
O6 Different manufacturing systems		●	●							●	○	○	●	○						●	●	○
O7 Electricity supply options	○	○			●	●		●				○					○			●	●	
O8 Connected systems	○	●	●					○	○		●	○		●						●		○
O9 Material and energy flow dynamics		●		●			●		○	○	○	●	○	○		○						●
O10 Holistic quantitative evaluation	○						○		○						●	●			○			○
O11 Configurable and applicable	○		○			○			○						○	●	○			●	●	
O12 Dynamic visualization		○		○			○								○	●		○				○
O13 Application structure and guidelines	○		○		○	○			○							○	○	○	○	●	●	

Further, a second control loop is highlighted, which performs execution-related control of the manufacturing system and its energy demand. The whole manufacturing system (comprising the management system and the performance system) has different inputs (e.g., materials and energy) and outputs (e.g., products and emissions), while overall reference objectives guide the production management, and external disturbances influence the manufacturing system. An information interface is used for data exchange with outside entities.

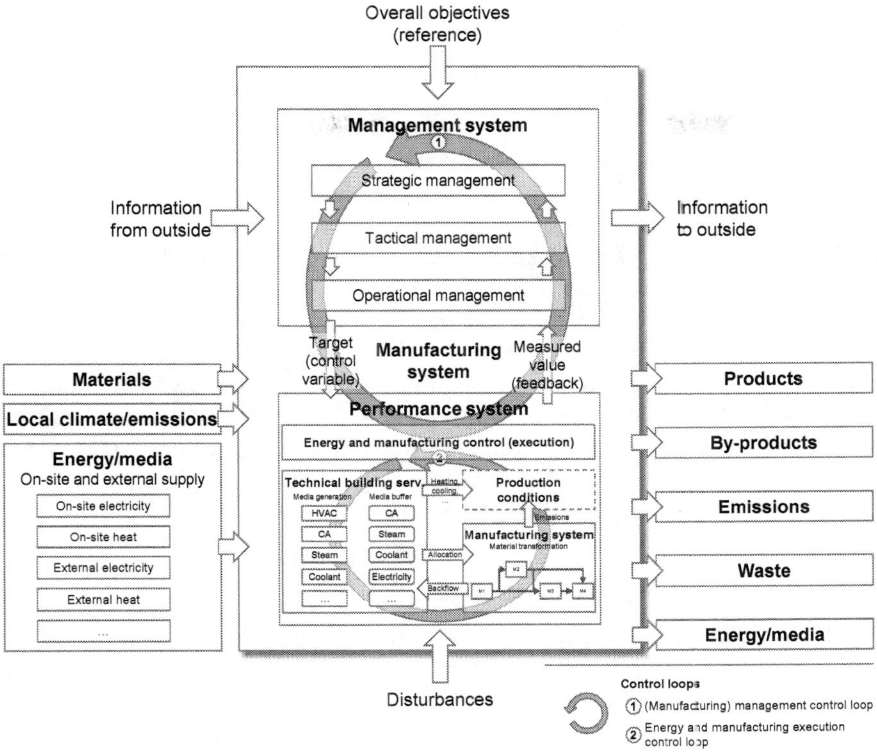

Fig. 4.3 Concept development starting point: combining the production management control cycle (Fig. 2.4) with the holistic factory understanding (Fig. 2.3) (own illustration based on referenced figures and their references, compare also to Thiede 2012; Herrmann et al. 2014 and Sproedt et al. 2015)

Within the developed concept, different interacting control cycles (management and execution) are used to implement and improve energy flexibility of a manufacturing system. In addition to the described theory outlined in Chap. 2 and Fig. 4.3, elements from the modeling approach of dynamic and interdependent energy demand of manufacturing systems described in Thiede (2012) influenced modeling of the manufacturing system. Further, the holistic view of material and energy flows and their quantification within a factory as described in Posselt (2016) provided input for detailing the manufacturing system material and energy flow behavior (compare also to Sect. 2.1.3).

4.2.1 Concept Structure

Based on the idea to understand manufacturing management and execution as one or more interlinked control cycles (compare to Fig. 4.3), the suggested concept is based on multiple linked feedback and control cycles. The proposed planning and improvement framework and its real-time control and system behavior model concept can be found in Fig. 4.4. The concept is split into two parts: first (upper part in Fig. 4.4), a planning and improvement framework is introduced, which reflects the different managerial planning horizons and thus includes the manufacturing management control loop (control loop one in Fig. 4.4, derived from the management control loop in Fig. 4.3). As a second part, the manufacturing system (performance system), based on the different elements from the holistic factory understanding, is detailed as specific methods and tools are required to successfully achieve and improve energy flexibility of a manufacturing system. The energy and manufacturing execution control loop from Fig. 4.3 is split into a central (labeled loop two) and decentralized (loop three) energy and manufacturing execution control loop within the concept.

Planning and Improvement Framework

As an overarching structure, a managerial planning framework is introduced, which encompasses the idea of splitting production management into different planning horizons. Five dedicated elements (I–V) are introduced and described in the following:

I Vision/Mission and Company Goals, Incl. Vision of Energy Flexible Factories

Beginning with the upper part of Fig. 4.4, the starting point consists of the overall (company) vision, mission, and objectives, including the vision for energy flexible factories that acts as an overarching frame for all other parts. Rather than including a specific target such as an energy-autarkic factory or integration of VRE into the factory, the term *vision of energy flexible factories* is used to account for the variety of different targets that might be included in and/or enabled by energy flexible factories. Based on the idea that an overall vision of a company guides all company planning tasks, energy flexibility should be reflected within this vision to ensure that it becomes an integral part of the company's policies (Wiendahl 2014). Within Fig. 4.3, the energy flexibility vision and mission provides overall objectives as reference for the (wider) manufacturing management control circle.

II Planning and System Configuration

Based on the vision and mission as reference, strategic, tactical, and operational (energy flexibility) planning is performed and related specific objectives are formulated (upper box of "II Planning and system configuration," corresponding to strategic, tactical, and operational management) to achieve the target state outlined within the vision and mission. Long- to short-term objectives are detailed (e.g., increased utilization of on-site generated VRE by 5 % within the next two years).

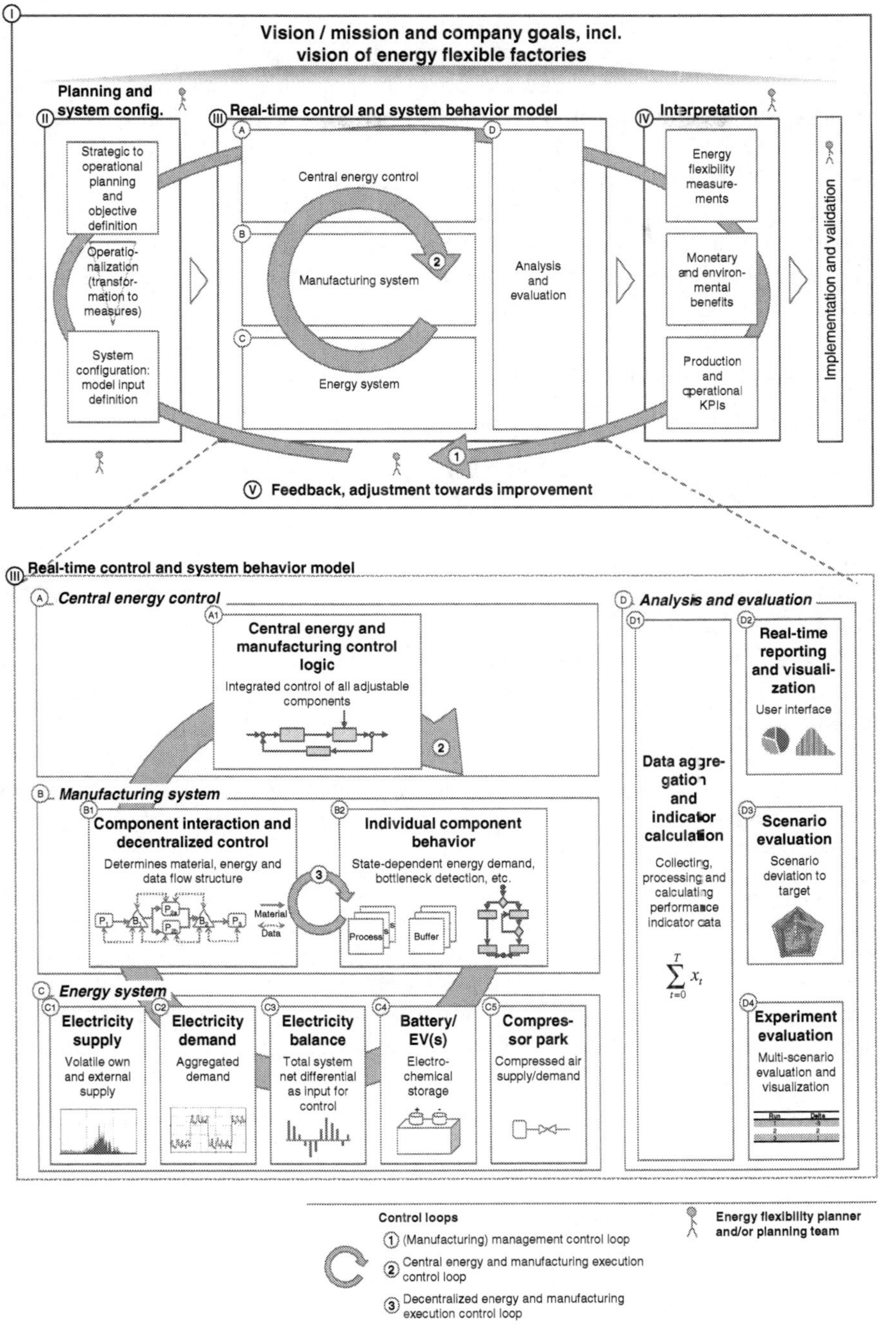

Fig. 4.4 Energy flexibility planning and improvement framework (*top*) and energy flexibility real-time control and system behavior model concept (*bottom*, own illustration)

These objectives are then operationalized by transforming managerial objectives into specific measures (e.g., increasing embodied energy storage in products or deploying battery storage, arrow between the two boxes in II). These measures are then used to define a system configuration (e.g., increase intermediate product buffer capacities by 20 % or install a stationary battery with 100 kWh capacity). This specific system configuration is used as input for the system behavior model in order to test impact and effectiveness of measures (lower box of "II Planning and system configuration" in Fig. 4.4). Referring to Fig. 4.3, described elements are reflected in the different planning horizons from the management system. However, strategic, tactical, and operational management goals and objectives, derived measures and system configurations are specifically formulated for a performance system model to allow testing and evaluation of estimated impacts before implementation.

III Real-Time Control and System Behavior Model

The quantitative, dynamic real-time control and system behavior model is then used to quantify the influence of previously described measures on selected (energy flexibility) indicators (e.g., direct demand of on-site generated electricity as a share of total electricity demand). Instead of a real system (performance system in Fig. 4.3), a model is proposed to evaluate measures before implementation in a real system. The model is based on the holistic understanding of a factory, with a focus on material and energy flows. The system behavior model forms the main part of proposed concept as estimating the manufacturing system behavior and applying an integrated energy and manufacturing execution control (central and decentralized control loops two and three in Fig. 4.4) are complex and dynamics tasks which need detailed attention. Further, supplying a comprehensive and detailed model also aims at developing specific measures and tools towards flexible energy demand of manufacturing systems as outlined in the initial research question.

IV Interpretation

In a forth step, obtained results from the system behavior model are used for detailed interpretation to determine target achievement towards energy flexibility objectives and their impact on further performance measures (measures related to, e.g., impact on the manufacturing system operation, economic, and environmental impact).

Additional analysis to already performed calculations within the system model as well as visualization to ease interpretation might be necessary. Referring to Fig. 4.3, this step is centered around providing feedback (measured values) from the performance system to the management system.

V Feedback, Adjustment Towards Improvement

In order to close the management control loop (control loop one in Fig. 4.4, compare also to management control loop in Fig. 4.3), the last step aims at determining appropriate actions and required changes (e.g., further increase of product buffer capacities instead of installing a battery) to improve target achievement of energy flexibility and related (manufacturing) system operational, environmental, and economical goals.

This step also accounts for potential disturbances from outside which require adjustment of initial goals and/or which impact results and thus need to be reflected within the different concept elements (sequential order of described elements I to V within the management control cycle, further highlighted in an application cycle described at the end of this chapter).

In summary, if desired results are achieved, defined measures need to be implemented in the real (performance) system. Further detail on implementation will also be provided in the application cycle at the end of this chapter.

As mentioned before, the real-time control and system behavior model is the central element of proposed concept due to its organizational, dynamic, and technical complexity and is thus described in the following.

Real-Time Control and System Behavior Model Concept

The real-time control and system behavior model concept is detailed in the lower part of Fig. 4.4. It consists of the four elements *Central energy control*, *Manufacturing system*, *Energy system*, and *Analysis and evaluation*. The central energy and manufacturing control (A/A1) aims at determining control signals for the manufacturing system to control energy demand. As such, it forms the integral element of the concept's second control circle, which is a real-time (execution-related) energy control of manufacturing systems. The manufacturing system (B) itself consists of two elements. The first (sub-)element (B1) determines dynamic component interaction, including material and energy flows, which are based on the second (sub)element (B2), individually acting agents that reflect behavior of all interacting system elements. A feedback loop and decentralized control influences system and agents behavior dependent on individual agents' actions and system state, e.g., to achieve constant throughput. The energy system (C) determines electricity supply from external and on-site sources (C1), demand from from the manufacturing system and connected elements (C2) and consequently the total supply/demand balance (C3). Resulting electricity surplus or shortage is then used as input for a connected battery and/or EV(s) (C4). Further, additional energy-centered system elements are integrated. For the purpose of this concept, compressed air produced in a compressor park (C5) is chosen to demonstrate integration of other energy carriers than electricity. The forth general system element, analysis and evaluation (D), has the task of collecting, processing and calculating performance indicator data (D1) and to report results. Reporting includes real-time visualization of indicators to study system behavior (D2). A special focus is set on multi-indicator scenario evaluation (D3) and multi-scenario experiment evaluation (D4) to quantitatively evaluate system performance and differences between different system structures and/or changing input parameters (based on hypotheses from energy flexibility objectives).

Target user of the concept is a person or team which resides in the broader production management function (labeled "energy flexibility planner and/or planning teams"). The energy flexibility planner derives tactical and operational energy flexibility goals and related measures to achieve these goals, uses the detailed behavior model to evaluate the effectiveness of defined measures and is responsible for results interpretation. Furthermore, the planner adjusts measures to improve target (energy flexibility objective) achievement. Finally, when desired results are obtained, the planner implements and validates measures.

Within the following sections, the real-time control and system behavior model's elements are described in detail to enable an outside person to reproduce the model. Further, the initially stated research question explicitly aims at developing specific measures and tools, which is targeted with describing the energy flexibility model and especially control in detail.

Considering the general definition of a model as "[..] an image of a sector of reality which is created in order to accomplish a given task" (Banerjee 2014, p. vii), the starting point is defining the purpose of a desired model. As mentioned before, proposed model aims at testing and validating hypotheses under a real-time control approach to integrate VRE into a manufacturing system and, related to this objective, to evaluate the system behavior. Banerjee classifies models into (Banerjee 2014, p. vii f.):

- Verbal descriptions
- Structured lists of measurements
- Maps of various kinds
- Real models
- Mathematical models

Within mathematical models, a further classification into dynamic models, geometric models, and classifying models is proposed. As the purpose of proposed concept is to evaluate behavior of a dynamic system, a mathematical model, in particular a dynamic model, is developed. A central component of mathematical models is, among others, a set of equations. For the case of dynamic models, differential equations are of particular interest. Therefore, the following sections reflect a mathematical description of proposed model, utilizing (differential) equations.

Model time is denoted as $t = 1, \ldots, T$. Total number of system elements that never enter or exit the system within model time (e.g., processes, buffers, compressors) is denoted as E, while $e = 1, \ldots, E$ is used as subscript for individual elements. The description starts with the energy system to provide necessary background for other model elements.

4.3 Energy System

The energy system consists of the five elements electricity supply, demand, balance, and related infrastructure, namely a battery and/or EV(s) and the compressor park (Fig. 4.5). The electricity supply module determines available time-dependent electricity from various sources (e.g., wind generation or grid supply). The electricity demand module dynamically aggregates electricity demand from all manufacturing system elements and connected auxiliary and infrastructure system elements (e.g., compressors). Both values, electricity supply and demand, are then used to determine the resulting balance, which is used as reference input for the central electricity

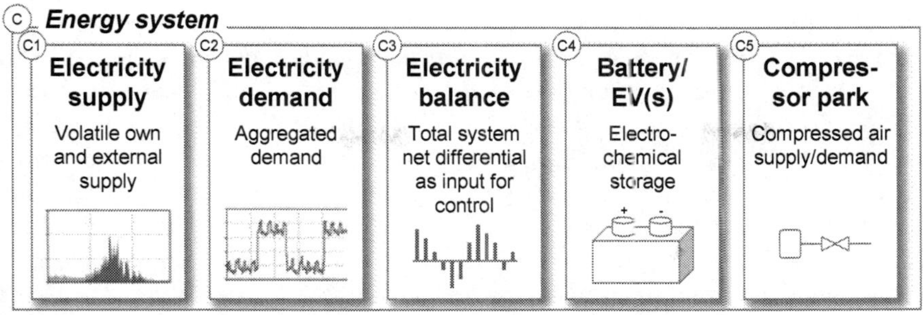

Fig. 4.5 Energy system elements (compare also to Fig. 4.4, own illustration)

control logic. Finally, a battery and/or EV(s) utilize the resulting electricity supply and demand balance to determine charging and discharging actions with the goal to reduce demand and supply mismatches. The compressor park as part of the auxiliary/infrastructure system is allocated to the energy system as CA is an energy carrier utilized in manufacturing processes and CA can be used to store energy and thus be controlled under energy flexibility goals.

4.3.1 Electricity Supply

Electricity supply can be ensured by deploying different supply sources. For the purpose of this concept, it is assumed that a set of VRE supply sources can be deployed on-site. These supply sources offer a supply that varies over time, i.e., total energy output is limited and changes over time. All on-site (own) generated electricity supply El_t^{own} at time t is calculated by adding time-dependent supply from individual supply sources:

$$El_t^{own} = \sum_{os=1}^{OS} El_{os,t}^{own}, \tag{4.1}$$

where $os = 1, \ldots, OS$ is used for indexing all individual electricity supply sources (i.e., wind turbines, solar panels/arrays or CHP plants).

Aside from own supply, electricity can be drawn from a connected (public) grid. Current grid demand is denoted as El_t^{grid}. Total electricity supply $El_t^{total} = El_t^{own} + El_t^{grid}$ is calculated as the sum of on-site generated electricity and electricity drawn from the grid. Note that all supply sources, own and grid supply, are indexed by $s = 1, \ldots, S$, and $\phi_{s,t}$ denotes the current supply share of supply source s (in percent, with $\sum_{s=1}^{S} \phi_{s,t} = 1$).

Above general description of supply structure requires a specific time series for own electricity supply to be included in the model. The time series can either be included from a physical model (i.e., electromechanical wind turbine model), an empirical model (i.e., regression of a solar panel's time series output data) or from measured values (i.e., recording of electricity output data from an existing (black box) VRE system). For examples of mathematical/physical models, see, e.g., Deshmukh and Deshmukh (2008), Zhang et al. (2013), or Yang et al. (2009). For a general overview and summary on energy models, see, e.g., Jebaraj and Iniyan (2006). Although any data source providing time-dependent electricity generation values, also with variable resolution, can be connected to the model, acquiring variable electricity supply data time series is not the main focus of this concept and is therefore not further detailed.

4.3.2 Electricity Demand

Electricity demand of the overall system needs to be dynamically calculated. Moreover, every change in the system state (e.g., state change of a production process, reduced compressor output) requires reevaluating electricity demand of the system to avoid inaccuracies. As such, a continuous time representation is required. Total electricity demand $El^{dem.}_{sys.,t}$ (in watts) of the system at time t is the sum of all system elements' individual electricity demand $El^{dem.}_{e,t}$ (in watts):

$$El^{dem.}_{sys.,t} = \sum_{e=1}^{E} El^{dem.}_{e,t}. \tag{4.2}$$

Details on determining individual electricity demand are provided within following sections on individual elements behavior.

Cumulative electricity demand $CEl^{dem.}_{sys.,t}$ is calculated by integrating system electricity demand at time t over time:

$$CEl^{dem.}_{sys.,t} = \int_0^t El^{dem.}_{sys.,t}\, dt \tag{4.3}$$

and denotes the total electricity demand of the system until time t (in watt-hours). Further cumulative/total electricity demand values (i.e., individual processes or for a specific energy source such as solar electricity demand) can be obtained similar to Eq. 4.3 and therefore detailed equations are not provided in the following.

4.3.3 Electricity Balance

Total electricity demand and supply flows need to balance at all times to ensure consistency in energy flows (no 'creation' of energy within system boundaries). The

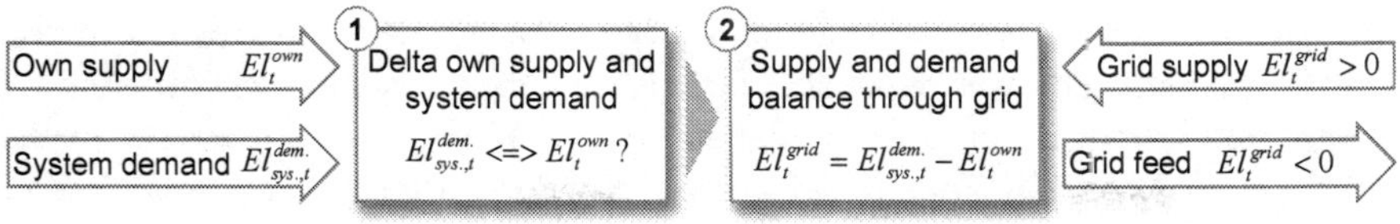

Fig. 4.6 Two step approach for establishing electricity supply and demand balance (own illustration)

proposed concept assumes that a set of own on-site supply sources with limited supply capabilities are available. Further, in the model's initial configuration, unlimited grid supply is assumed to be available. In order to maximize demand of own electricity, electricity is only drawn from the grid if own supply is not sufficient to meet demand. Consequently, total system demand $El_{sys.,t}^{dem.}$ and own supply determine total grid supply El_t^{grid}:

$$El_t^{grid} = El_{sys.,t}^{dem.} - El_t^{own}. \tag{4.4}$$

Above relation is visualized in Fig. 4.6. Two characteristics of the equation are important to note: first, the equation assumes that El_t^{grid} can be negative if $El_t^{own} > El_{sys.,t}^{dem.}$. This relation implicitly assumes that own electricity can be fed back into the grid. Second, the integral over time, similar to Eq. 4.3, for all t for which $El_t^{grid} > 0$ yields total/cumulative external electrical energy demand. This indicator is, for example, important to determine electric energy cost (common billing value) and magnitude of environmental impact if values for specific (per Wh) environmental impact are available (e.g., amount (weight) of carbon emissions from fossil fuel fired power plants per Wh). Further, this value can be used as indicator for grid dependency, and change thereof due to on-site supply.

If more than one on-site supply source is available and electricity is fed to the grid due to oversupply, an allocation rule needs to be enacted to determine which supply source is fed to the grid. For simplicity, an economic-based rule is applied: the supply source with the highest feed-in compensation (e.g., Euro per kWh) is assumed to be ranked first for feed-in, followed by a decreasing feed-in compensation order. If no feed-in compensation values are available or compensation is equal, an equal share for each source is assumed. Nonetheless, any other feed-in rule (i.e., due to technical considerations) can be used.

The described basic model can be extended to a case of limited grid supply, i.e., to determine DR capabilities for a single event. An artificial limited supply source can be defined and included in the system similar to other limited (own) supply sources os (see also Eq. 4.1). For example, an own supply source can be defined which supplies a constant amount of power. Power is reduced immediately to a lower amount or zero and, after the assumed duration of the DR event, increased to its initial value. The magnitude of grid supply during this period indicates the DR capability of the system, i.e., if the DR event can be accommodated (if grid supply was zero throughout the whole DR time) or what the maximum power reduction for the given DR time is (DR event power reduction (negative) plus highest grid supply). Further, utilizing the resulting grid supply time series profile during the DR event, maximum DR power reduction for a given DR duration or maximum DR duration for a given DR power reduction can be evaluated.

4.3.4 Battery/EV(s)

As outlined in the specific requirements for a new concept (Sect. 4.1), decentralized energy storage should be included in the model. As batteries are a promising method to support integration of VRE (cf. Sect. 2.3.1), especially in the light of expected future increase in EV deployment, a stationary battery and/or connected EV(s) are considered here. One or more battery/EV module(s) can be deployed to mitigate supply and demand mismatches. As the concept explicitly aims to be modularly structured, the battery and/or EV(s) are optional.

Battery

Physical battery models have been studied extensively, also for use in renewable energy models to mitigate VRE fluctuations (see, e.g., Guasch and Silvestre 2003, Yang et al. 2007, 2009, Bajpai and Dash 2012, Bhandari et al. 2014, and Li et al. 2015, an application for stand-alone VRE supply with battery can be found in Brent and Rogers 2010, for a review on hybrid renewable energy (e.g., VRE-battery) system design, control, simulation and optimization approaches, methods and tools see, e.g., Bernal-Agustín and Dufo-López 2009, Zhou et al. 2010 and Bajpai and Dash 2012). Based on existing methods and models, a dynamic battery model is developed (Fig. 4.7). Connected EVs are assumed to behave similarly, with the exception that EVs might become temporarily unavailable, which is described after introduction of the (generic) battery model. Each battery included in the system is separately modeled and all batteries together form the battery system, with $b = 1, \ldots, B$ denoting the different batteries with total battery count B. The individual battery model is generic, i.e., not battery-type or technology specific and requires seven battery-specific parameters for configuration (one set for each included battery):

- $BC_{b,0}$: Initial battery capacity (in watt-hours)
- CR_b: Battery maximum charge rate (in watts)
- $Cyc_b^{80\,\%}$: Number of discharge cycles until 80 % initial capacity remains (in cycle-counts)
- DCR_b: Battery maximum discharge rate (in watts)
- η_b: Battery efficiency (in percent)
- ξ_b: Self-discharge factor (in percent)
- $SOC_b^{init.}$: Initial battery charge (in watt-hours)

Initial battery capacity $BC_{b,0}$ is assumed to decrease over time if the battery is charged and discharged, and consequently $BC_{b,t}$ is used to describe the battery's

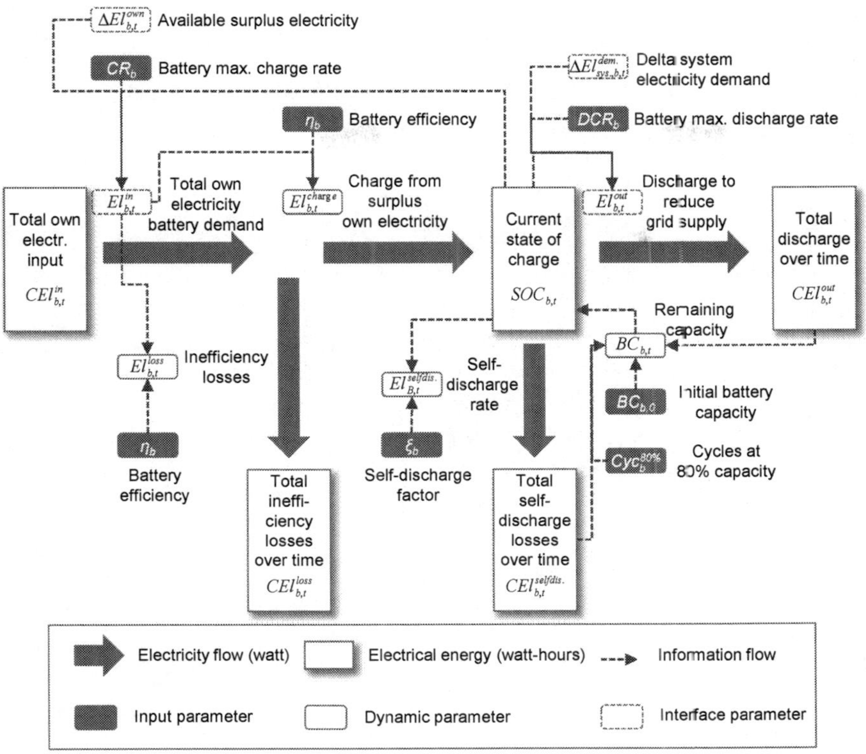

Fig. 4.7 Battery model structure (own illustration)

remaining capacity at t. Battery lifetime or lifespan is dependent upon several factors (i.e., (dis-)charge rate, depth of discharge), is highly nonlinear and thus complicated to model in detail (for a discussion of battery lifetime models see, e.g., Li et al. 2015). For simplicity, number of discharge cycles $Cyc_b^{80\,\%}$ until 80 % of initial capacity remains is chosen as input parameter, temporal aging is disregarded. One cycle is defined to be completed if the battery has been fully discharged. For example, discharging to 75 % capacity level and recharging to full level four times is equal to one cycle, as two times discharge to 50 % and corresponding twice recharging would be. This definition is also used by official organizations (e.g., United Nations 2009) and electronics manufacturers (e.g., Apple Inc. 2015). For further details on battery lifespan estimations, see, e.g., Andrea (2010).

Battery remaining capacity is assumed to decrease linear with number of discharge cycles. Total number of discharge cycles $Cyc_{b,t}$ is calculated by

$$Cyc_{b,t} = \frac{CEl_{b,t}^{out} + CEl_{b,t}^{selfdis.}}{BC_{B,0}} \tag{4.5}$$

where $CEl_{b,t}^{out}$ is the total electricity output of the battery over its lifetime and $CEl_{b,t}^{selfdis.}$ total electricity self-discharge losses. For simplicity and to obtain a closed form expression, initial battery capacity $BC_{b,0}$ is utilized as denominator to calculate $Cyc_{b,t}$ (and not $BC_{b,t}$). Remaining capacity is then calculated by

$$BC_{b,t} = BC_{b,0} \cdot \left[1 - \frac{0.2 \cdot Cyc_{b,t}}{Cyc_b^{80\,\%}} \right]. \tag{4.6}$$

After determining the battery's capacity, battery charge and discharge dynamics are modeled. Starting left in Fig. 4.7, total own electricity battery demand $El_{b,t}^{in}$ is determined by available own electricity which is not immediately demanded by the system and would thus be otherwise fed into the grid. For the case of more than one battery in the system, allocated available surplus electricity $\Delta El_{bnf,t}^{own}$ for battery bnf is determined according to the batteries charge rate in relation to total charge rate of all batteries which are not full:

$$\Delta El_{bnf,t}^{own} = \frac{CR_{bnf}}{\sum_{bnf=1}^{BNF} CR_{bnf}} \cdot \Delta El_t^{own} \ \forall bnf \in BNF, \tag{4.7}$$

where $BNF \in B$ denotes all batteries that are not fully charged, i.e., their state of charge is $SOC_{bnf,t} < BC_{bnf,t}$, and with

$$\Delta El_t^{own} = El_t^{own} - El_{sys.,t}^{dem.}. \tag{4.8}$$

For all batteries with $SOC_{b,t} = BC_{b,t}$, allocated available surplus electricity is zero.

Considering the battery's maximum charge rate CR_b, current state of charge $SOC_{b,t}$ and remaining capacity $BC_{b,t}$, total own electricity battery demand is

$$El_{b,t}^{in} = \begin{cases} CR_b & \text{if } \Delta El_{b,t}^{own} \geq CR_b \wedge \Delta El_{b,t}^{own} > 0 \wedge SOC_{b,t} < BC_{b,t} \\ \Delta El_{b,t}^{own} & \text{if } \Delta El_{b,t}^{own} < CR_b \wedge \Delta El_{b,t}^{own} > 0 \wedge SOC_{b,t} < BC_{b,t} \\ 0 & \text{if } \Delta El_{b,t}^{own} \leq 0. \end{cases} \tag{4.9}$$

In other words, the battery is charged with either its maximum charge rate or with its individual maximum available surplus electricity (whichever is lower), given that surplus electricity is available and that the battery is not full (which is already reflected in Eq. 4.7, but included here for completeness and for simplification if only one battery is modeled, i.e., Eq. 4.7 can then be omitted).

The battery is assumed to incur energetic (exergetic) 'losses' during charging and discharging, i.e., electricity supplied to the battery cannot be fully obtained back from the battery due to, e.g., heat losses as a result of internal electrical resistance and chemical reactions.[2] However, separately measuring charging and discharging efficiency is difficult. Consequently, only one efficiency measure (round-trip efficiency η_b) is introduced, which affects charging efficiency (similar to Guasch and Silvestre 2003, which is also used in Yang et al. 2007, 2009). Therefore, losses due to battery inefficiency are:

$$El_{b,t}^{loss} = (1 - \eta_b) \cdot El_{b,t}^{in} \tag{4.10}$$

and electricity charge flow into the battery

$$El_{b,t}^{charge} = \eta_b \cdot El_{b,t}^{in} = El_{b,t}^{in} - El_{b,t}^{loss}. \tag{4.11}$$

Consequently, discharging incurs no losses within proposed model.

The current state of charge $SOC_{b,t}$ of the battery is calculated by integrating battery charge, discharge, and self-discharge rates over time and adding initial charge $SOC_b^{init.}$ (if any):

$$SOC_{b,t} = \int_0^t \left[El_{b,t}^{charge} - El_{b,t}^{out} - El_{b,t}^{selfdis.} \right] dt + SOC_b^{init.}. \tag{4.12}$$

Self-discharge $El_{b,t}^{selfdis.}$ of the battery is assumed to be dependent on $SOC_{b,t}$ and a self-discharge factor ξ_b (in percent) (Guasch and Silvestre 2003; Yang et al. 2009):

$$El_{b,t}^{selfdis.} = \xi_b \cdot SOC_{b,t}. \tag{4.13}$$

For simplicity, ξ_b is assumed to remain constant over the battery lifespan.

Battery discharge to reduce grid supply $El_{b,t}^{out}$ depends on the battery's $SOC_{b,t}$, the battery's maximum discharge rate DCR_b and the amount of electricity demand exceeding own supply that is allocated to the battery. Similar to Eq. 4.7, allocated discharge amount $\Delta El_{sys.,bne,t}^{dem.}$ is

$$\Delta El_{sys.,bne,t}^{dem.} = \frac{DCR_{bne}}{\sum_{bne=1}^{BNE} DCR_{bne}} \cdot \Delta El_{sys.,t}^{dem.} \quad \forall bne \in BNE, \tag{4.14}$$

with $bne = 1, \ldots, BNE$ and $BNE \in B$ denoting all batteries which are not fully discharged (i.e., $SOC_b > 0$) and

$$\Delta El_{sys.,t}^{dem.} = El_{sys.,t}^{dem.} - El_t^{own} = -\Delta El_t^{own}. \tag{4.15}$$

[2] The term *losses* is used here to differentiate from a desired energy outflow from the battery, although losses are, with regards to the law of conservation of energy, one type of outflow from the system.

Similar to Eq. 4.9, battery discharge is calculated by

$$El^{out}_{b,t} = \begin{cases} DCR_b & \text{if } \Delta El^{dem.}_{sys.,b,t} \geq DCR_b \wedge \Delta El^{dem.}_{sys.,b,t} > 0 \wedge SOC_{b,t} > 0 \\ \Delta El^{dem.}_{sys.,b,t} & \text{if } \Delta El^{dem.}_{sys.,b,t} < DCR_b \wedge \Delta El^{dem.}_{sys.,b,t} > 0 \wedge SOC_{b,t} > 0 \\ 0 & \text{if } \Delta El^{dem.}_{sys.,b,t} \leq 0. \end{cases} \quad (4.16)$$

The battery discharge rate is equal to its maximum discharge rate or current excess allocated system demand (whichever is lower), given that own supply is not sufficient to satisfy demand and that the battery is not empty.

Referring to Fig. 4.7, cumulative electricity values CEl_t (in watt-hours) are calculated by integrating input/output values El_t over time (similar to Eq. 4.3). In summary, Eqs. 4.5–4.16 constitute the dynamic battery model.

Interfaces to other (submodels) are time-dependent electricity input ($El^{in}_{b,t}$) and output ($El^{out}_{b,t}$). If a battery is part of the overall system, Eq. 4.4 becomes

$$El^{grid}_t = El^{dem.}_{sys.,t} - El^{own}_t + El^{in}_{bat.,t} - El^{out}_{bat.,t}, \quad (4.17)$$

with $El^{in}_{bat.,t} = \sum_{b=1}^{B} El^{in}_{b,t}$ and $El^{out}_{bat.,t} = \sum_{b=1}^{B} El^{out}_{b,t}$. Electricity feed into the grid is reduced through battery charging and electricity supply from the grid is reduced through battery discharging. Consequently, grid dependency is reduced and on-site generated electricity is utilized locally, adjusted for conversion inefficiencies.

For the case of multiple own supply sources, stored electricity needs to be separately tracked by supply source for further (later) allocation into embodied electricity (e.g., in products or compressed air). In order to track how much electricity from a given source is stored in a battery at any time, an individual electricity flow model for each supply source needs to be set up per battery. However, the combined battery model (all supply sources) and subsequently their combination into a battery system is utilized as interface to other modules.

As outlined above, if more than one battery is attached to the system, each battery needs to be modeled separately to ensure dynamic consistency unless multiple batteries are connected with each other and appear as a single system, which implies that physical values such as charge rate and capacity can be combined and still reflect individual battery characteristics. However, if batteries individual dynamic behavior (charging, discharging) needs to be evaluated separately (e.g., due to different charging/discharging/capacity values, individual battery failure or removal as part of an electric mobility system connected to the manufacturing/VRE system), an individual battery model as described above for each battery needs to be set up. The interface to the overall system is then the sum of the batteries' individual values, i.e., total charge rate is the sum of all charge rates. Maximum charge rate is then defined by the individual batteries' maximum charge rates, state of charge and availability if batteries can be dynamically removed. Actual battery charge and discharge rates (if less than the sum of all batteries' maximum rates) are then allocated proportionally to the batteries' individual maximum charge rates (i.e., two batteries with 1 kW and

3 kW maximum charge rate and surplus own electricity supply of 2 kW would result in charging the first battery with 0.5 kW and the second battery with 1.5 kW).

Electric Vehicles

The dynamic behavior of EVs is assumed to be similar to batteries (charge and discharge behavior), with the exception that (a) each EV needs to be modeled as a separate battery (module) and (b) that EVs might become unavailable for charging and discharging, and (c) that an unavailable EV has a reduced charge upon return as the EV's battery is discharged during driving of the vehicle.

Different EV modules can be set up using previously described logic for multiple (stationary) batteries. If an EV is removed from the system, its current charge becomes unavailable for electricity supply to the manufacturing system. Further, surplus electricity from VRE cannot be used to charge the EV and needs to be fed into the grid (if remaining batteries and/or EVs cannot be charged as they are either unavailable, fully charged or already charged with their respective maximum rate).

The main purpose of EVs is their use for transportation and as such, transportation requirements should not be impacted by using EVs as intermittent energy storage option for VRE integration. Therefore, a logic is implemented which determines the remaining time until an EV is required for transportation and thus removed from its charging station (this logic assumes that a vehicle driving schedule is upfront known, e.g., for a case where EVs are scheduled as part of a company vehicle fleet). The EV logic then determines if a given vehicle can only be charged but not discharged, which depends on planned driving distance and thus required battery charge, the EVs current SOC, maximum charge rate and remaining time until removal. If the vehicle charging logic determines that the time until removal is equal to the minimum time required for reaching the needed SOC, the vehicle is charged with its maximum rate with electricity from the grid or VRE (if surplus VRE is available), but not discharged anymore. Upon return of the vehicle (reentering the modeled system), the vehicle's SOC is reduced by the amount required for the assumed driving distance, connected back to the local electricity grid and thus again available to the system.

4.3.5 Compressor Park

The compressor park consists of several individual compressors and a connected storage tank. CA distribution, e.g., flow dynamics, is not modeled in detail. However, additional volume from piping needs to be included in overall CA storage volume (for simplicity, CA tank size and system volume is referred to as CA system storage volume in the following, denoted as $CASV_{sys.}$ (in, e.g., m^3)). Individual compressor behavior and energy demand will be detailed after describing the compressor park logic.

Compressor Park Logic

The compressor park supply and demand structure can be found in Fig. 4.8 (modeling based on Ruppelt 2003, Fraunhofer ISI 2003 and Müller et al. 2009). Air is

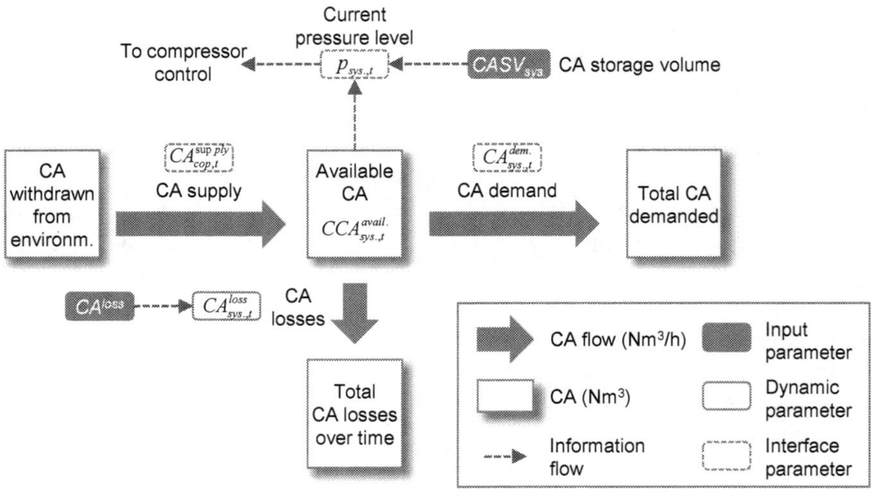

Fig. 4.8 Compressor park supply and demand structure (own illustration)

withdrawn from the environment and compressed by switched on compressors with a rate $CA^{supply}_{cop,t}$ (e.g., in Nm³/min, *cop* refers to compressor park) into the CA storage volume. Total CA system losses are denoted as $CA^{loss}_{sys.,t}$ (e.g., in Nm³/min) and summarize all CA losses within the system, e.g., due to leaks. Further, (manufacturing system) CA demand is denoted as $CA^{dem.}_{sys.,t}$. Available CA at time t is denoted as $CCA^{avail.}_{sys.,t}$ (cumulative CA, in Nm³).

Total CA supply is calculated by adding individual compressor supply

$$CA^{supply}_{cop,t} = \sum_{cs=1}^{CS} CA_{cs,t}, \tag{4.18}$$

with compressor index $cs = 1, \ldots, CS$.

For simplicity, CA system losses are assumed to be measured or estimated and therefore $CA^{loss}_{sys.,t}$ is equal to a constant input parameter CA^{loss}.

Total CA demand at time t is the sum of all system elements' CA demand and thus

$$CA^{dem.}_{sys.,t} = \sum_{e=1}^{E} CA^{dem.}_{e,t}. \tag{4.19}$$

Total available CA at time t is the integral of in- and outflows into the storage until time t plus CA initial fill level $CA^{init.}_{sys.}$:

$$CCA^{avail.}_{sys.,t} = \int_0^t \left[CA^{supply}_{cop,t} - CA^{loss}_{sys.,t} - CA^{dem.}_{sys.,t}\right] dt + CA^{init.}_{sys.}. \tag{4.20}$$

Assuming that equipment requires a certain minimum pressure, compressor control utilizes system pressure p_t as measured input (Ruppelt 2003). Assuming that CA behaves like an ideal gas and assuming constant temperature, the ideal gas law can be used to determine system pressure (Ruppelt 2003; Fraunhofer ISI 2003). With the ideal gas law (volume V in m^3 and constant temperature)

$$p_1 \cdot V_1 = p_2 \cdot V_2 \tag{4.21}$$

and an ambient pressure of one bar, CA system pressure at time t is

$$p_{sys.,t} = \frac{CCA_{sys.,t}^{avail.}}{CASV}. \tag{4.22}$$

Note that above equation is used to determine current system pressure. For compression (compressors), measured (state-dependent) energy demand is utilized within the model. Thus, low specific energy requirements for compression assuming an isothermal process are not utilized for determining compressor energy demand to avoid underestimating energy requirements.

Two compressor control strategies are introduced (Fig. 4.9). The first strategy, summarized under *no control*, denotes a full load-idle-stop/start control, which is a common control for compressors without variable speed drives (Fraunhofer ISI 2003, Ruppelt 2003, Bierbaum and Hütter 2004, also used in Thiede 2012). If system pressure reaches a certain minimum pressure p_{min}, compressors receive a signal to start compression. If system pressure reaches p_{max}, compressors receive a signal to stop compression. Depending on CA storage volume and CA demand, compressors either switch from idling to off (after passing their set idle waiting times) or, if system pressure reaches p_{min} before waiting time has passed, back to compression. Resulting pressure profile rises and falls between p_{min} and p_{max}, with changing slope if CA demand changes. The purpose of this strategy is to ensure adequate system pressure, while realizing limited compressor switches to prevent wear-out. Assuming that CA demand and compressor CA production capacity are exogenously given,

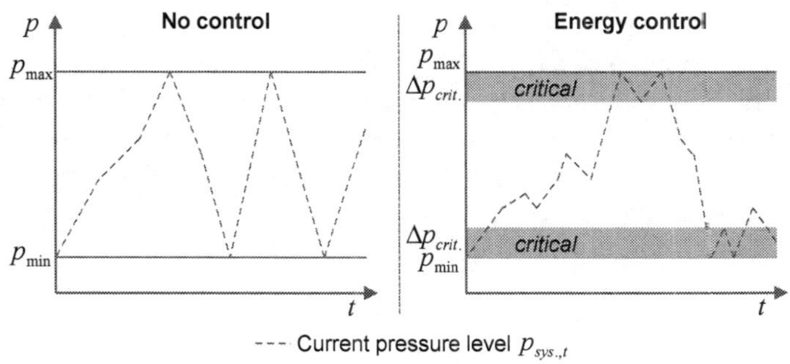

Fig. 4.9 Compressor park pressure control (own illustration)

free variables to influence compressor switches are compressor idle waiting times, system operating pressure band (p_{min} to p_{max}) and CA storage volume. Further, a cascading control can be enacted to optimize compressor switches and load, which is characterized by different pressure operating bands per compressor (for determining CA storage volume and further details on cascading controls see, e.g., Ruppelt 2003; IHK Nürnberg für Mittelfranken 2012, or Fraunhofer ISI 2003).

The second control strategy is applied when the central energy control to increase energy flexibility is enacted (right side of Fig. 4.9). Three different control states are introduced, dependent on system pressure p_t:

1 **Energy flexibility control**: If CA supply/system pressure is not determined critical (cf. Fig. 4.9), compressor control is left to the central electricity control (described in later sections). System pressure is not used to determine control signals for compressors until system pressure enters a critical state.
2 **Low pressure criticality**: If critical low pressure is detected ($p_t = p_{min}$), compressor control is, in a first step, only allowed by the compressor park control. Overall target of the control is to switch compressors on or off to achieve a CA production which is sufficient, but not significantly higher than CA demand. This logic is applied, assuming an unchanged CA demand, to avoid unnecessary electricity demand increase under energy flexibility control (the central electricity control has tried to achieve a lower compressor electricity demand, which resulted in a critical low pressure event, and therefore additional electricity demand is likely to result in a worse fit with central electricity control targets). First, all compressors which are adjustable (i.e., not out of order) are determined. Target CA output $CA^{target}_{cop,t}$ is set to equal CA demand and losses, adjusted by a factor ι_{CA} (e.g., in percent) to prevent the system to drop below critical pressure due to sudden increase in CA demand and potential delayed (additional) compressor start. Consequently, target CA production is

$$CA^{target}_{cop,t} = \left(CA^{dem.}_{sys.,t} + CA^{loss}_{sys.,t}\right) \cdot (1 + \iota_{CA}). \tag{4.23}$$

In order to determine the closest possible CA supply fit, all switch on/off combinations for adjustable compressors are determined. Let $csa = 1, \ldots, CSA$ be the index for all adjustable compressors ($CSA \in CS$), then total number of possible on/off combinations NC is $NC = 2^{CSA}$, with index $nc = 1, \ldots, NC$. Corresponding possible CA supply of a given combination is CA_{nc}. As the method aims at determining the combination which avoids further pressure drop, the optimal combination nc_o is chosen which yields the smallest positive delta between CA supply and demand:

$$\begin{aligned} &CA_{nc_o} - CA^{target}_{cop,t} \le CA_{nc} - CA^{target}_{cop,t} \quad \forall nc \neq nc_o \\ &CA_{nc_o} - CA^{target}_{cop,t} \ge 0. \end{aligned} \tag{4.24}$$

After setting production and switch off signals according to cs_o, the central electricity control is only allowed to switch compressors on, but not to switch compressors off during low pressure critical state. Compressor park critical state due to minimum pressure event lasts until a set pressure threshold has been passed ($p_{min} + \Delta p_{crit.} \leq p_{sys.,t}$, with $\Delta p_{crit.}$ (in bar) as additional input parameter). This ensures that the system reaches a state where some freedom for central electricity control exists before allowing central electricity control to reduce CA supply below demand. If CA demand or supply (i.e., due to compressor failure) changes during a criticality event, rescheduling becomes necessary and compressor on/off control signals according to Eq. 4.24 are newly determined and applied.

3 **High pressure criticality**: High pressure criticality is similar to low pressure criticality, but triggered if $p_{sys.,t}$ is equal to p_{max}. The goal of compressor park control is then to determine control signals for compressors to achieve a slightly lower CA supply than demand to reduce current pressure level. Target CA supply then becomes

$$CA^{target}_{cop,t} = \left(CA^{dem.}_{sys.,t} + CA^{loss}_{sys.,t}\right) \tag{4.25}$$

with no adjustment factor as a high pressure event is usually caused by excessive compressor CA generation as a result of high on-site electricity availability under energy control, and thus demand thereof is not unnecessarily curtailed. Further, higher system pressure due to sudden changes in demand are less relevant as short-term slightly higher pressure levels can normally be accommodated by system elements until compressor output is reduced (set to idle immediately or within seconds). If own electricity supply is reduced, energy control has, due to available CA, still the option to reduce electricity demand through compressor control. Equation 4.24 then becomes

$$\begin{aligned} &CA^{target}_{cop,t} - CA_{nc_o} \leq CA^{target}_{cop,t} - CA_{nc} \;\; \forall nc \neq nc_o \\ &CA^{target}_{cop,t} - CA_{nc_o} \geq 0 \end{aligned} \tag{4.26}$$

to determine the compressor switch on/off combination which yields the smallest negative delta between CA supply and demand plus losses.

Total number of compressor switches is dependent on CA system design, CA demand pattern and, for the case of enabled energy control, also on dynamic control signals. Consequently, compressor switches need to be closely monitored. If a certain hourly switch count is exceeded, measures need to be adopted to reduce switch count (e.g., avoid switching off the compressor and control pressure via remaining compressors or alter system design (e.g., larger CA tank size), if possible).

Tracking supply source-dependent embodied electricity in compressed air (i.e., the energy required to produce the amount of available compressed air, similar to embodied electricity and energy in products, compare also to Sect. 4.4.1) is essential to allocate embodied energy and electricity in products back to individual supply sources (i.e., CA used to process a product adds a certain amount of embodied electricity to the product, depending on the electricity that was required to supply the

CA) and to remove model start- and end-state influences on indicators. Considering the current supply share of all electricity supply sources,

$$\phi_{s,t}^{supply} = \frac{El_{s,t}}{\sum_{s=1}^{S} El_{s,t}}, \tag{4.27}$$

and total electricity demand of all compressors $El_{cop,t}^{total} = \sum_{cs=1}^{CS} El_{cs,t}$, added embodied electricity in CA at time t from supply source s is $\phi_{s,t}^{supply} \cdot El_{cop,t}^{total}$. Let $\phi_{s,t}^{CCA}$ (in percent) denote the current share of embodied electricity from a given source in available CA, $EICA_t$ the electric energy intensity of currently available compressed air (e.g., in Wh/Nm^3), embodied electricity in compressed air from supply source s at time t is

$$CEl_{s,t}^{CCA} = \int_0^t \left[\left(\phi_{s,t}^{supply} \cdot El_{cop,t}^{total} \right) - \left(\phi_{s,t}^{CCA} \cdot EICA_t \cdot \left(CA_{sys.,t}^{loss} + CA_{sys.,t}^{dem.} \right) \right) \right] dt. \tag{4.28}$$

Above equation integrates inflows from a given supply source into embodied electricity (which is dependent on supply source share during CA production and compressor idling) and subtracts the integrated embodied electricity outflows (which depend on CA embodied electricity source share and CA demand plus losses).

The share of embodied electricity in CA from a given supply source, $\phi_{s,t}^{CCA}$, is its relative share of embodied electricity from all supply sources

$$\phi_{s,t}^{CCA} = \frac{CEl_{s,t}^{CCA}}{\sum_{s=1}^{S} CEl_{s,t}^{CCA}} \tag{4.29}$$

and embodied electricity per CA is calculated from current total embodied electricity from all sources and available CA:

$$EICA_t = \frac{\sum_{s=1}^{S} CEl_{s,t}^{CCA}}{CCA_{cop,t}^{avail.}}. \tag{4.30}$$

With above equations and known CA demand from a given process, embodied energy from a specific source added to a product which is converted within the process is calculated by multiplying the current electricity intensity of compressed air ($EICA_t$) with the current embodied energy share of available compressed air by source ($\phi_{s,t}^{CCA}$) and the process current compressed air demand ($CA_{n,t}^{dem.}$).

Compressor Model

Compressor behavior is modeled similar to manufacturing process behavior, which will be further detailed in the following Sect. 4.4.1. A state-based logic to determine compressor output and energy demand is introduced (Ruppelt 2003). Compressors are assumed to know the following operating states:

- **Off:** The compressor is switched off, produces no CA and requires El_{cs}^{off} electricity (in most cases assumed to be zero).
- **Switching on:** The compressor is switched on, during which no CA is produced but $El_{cs}^{turn\ on}$ is demanded. The compressor enters idle after required switching time has passed.
- **Switching off:** Similar to switching on, a compressor enters this state from idle and switches to off after an (optional) turn-off time has passed (CA production is zero, electricity demand $El_{cs}^{turn\ off}$).
- **Idle:** The compressor is ready for production, i.e., the motor is switched on but without load. CA production is zero and electricity demand El_{cs}^{idle}.
- **Produce:** CA is produced with rate $CA_{cs}^{prod.}$ (e.g., in Nm^3/min), corresponding electricity demand is $El_{cs}^{prod.}$. Can only be entered from idle state and enters into idle if production is set to be switched off.
- **Out of order:** Denotes random failure behavior similar to manufacturing processes. Can be entered from every state except *off*, electricity demand and CA production are zero, parameters to describe failure and repair time are MTBF and MTTR.

Compressor behavior is similar to a binary process (described in the next section), i.e., either the compressor runs at (full) capacity or is idling/switched off (e.g., no variable speed drives are considered). Current compressor output is $CA_{cs,t}$ and electricity demand $El_{cs,t}$.

Compressor operating control is realized within the compressor park module, i.e., individual compressors receive control signals from the central compressor park control or central energy control (depending on current energy control scenario/CA system pressure state). Corresponding electricity demand dynamics in relation to control signals are visualized in Fig. 4.10. A typical discontinuous compressor control is chosen (full load-idle-stop/start): a given compressor is not switched off

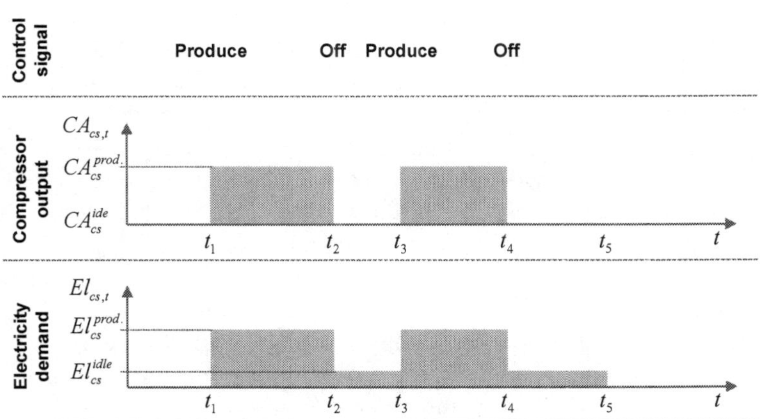

Fig. 4.10 Compressor CA output and electricity demand dynamics depending on control signals (own illustration)

Table 4.2 Maximum allowed compressor switches per hour depending on compressor size. Left: large compressors according to Müller et al. (2009), middle: large compressors according to IHK Nürnberg für Mittelfranken (2012), right: small compressors according to Ruppelt (2003)

Compr. power in kW	Max. switches per hour	Compr. power in kW	Max. switches per hour	Compr. power in kW	Max. switches per hour
200–250	5	250	4	15–22	12
110–160	10	110	8	7.5–11	15
65–90	15	30	15	4–5.5	20
30–55	20	7.5	30	1.5–3	25
11–22	25			0.37–1.1	30
4–7.5	30				

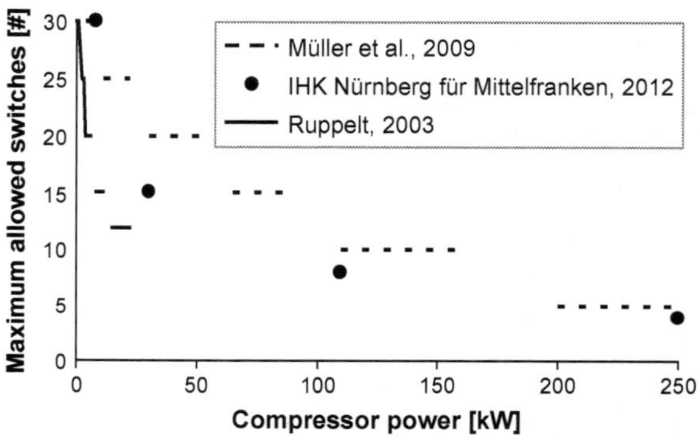

Fig. 4.11 Visualization of maximum compressor switches, based on Table 4.2 (own illustration)

immediately when CA production is switched off, but set into idle to avoid excessive motor switches and thus motor damage (Ruppelt 2003; Fraunhofer ISI 2003; IHK Nürnberg für Mittelfranken 2012; Bierbaum and Hütter 2004). If a compressor receives a control signal to be switched off (time t_2 and t_4 in Fig. 4.10), it enters into idle state, reduces CA output to zero, and an internal timer is started. If a preset wait time wt_{cs} (e.g., in seconds) has passed on no further control signal is received, the compressor motor is switched off (time t_5 in Fig. 4.10, with $t_5 - t_4 > wt_{cs}$). However, if the compressor receives a signal to resume CA production before waiting time has passed, the compressor can enter into production mode without delay and without increasing motor switch count (time t_3 with $t_3 - t_2 < wt_{cs}$ in Fig. 4.10). Compressor hourly switch count at time t is denoted as $Cc_{cs,t}$ (in switches per hour, rolling average calculation). Note that a longer waiting time wt_{cs} can result in less switches, but generally requires more electricity input as unproductive idle times are increased.

Table 4.2 and Fig. 4.11 show the maximum allowed number of compressor switches per hour as a function of compressor power. Depending on the literature source, maximum allowed switches can vary significantly from each other. This indicates that no general/strict rule for maximum allowed switches can be applied and individual operating instructions (manuals) of compressors should be consulted to obtain a case-specific value. Nonetheless, a lower number of switches is preferable. Control and operating strategies are discussed within the compressor park section. However, without energy flexibility control, CA system volume, operating pressure band, CA output, CA demand, and idle wait time of compressors are the main influencing factors for adjusting switch count.

4.4 Manufacturing System

The second main element of proposed model is the manufacturing system submodel. Its individual elements, component interaction, and decentralized control and individual component behavior (Fig. 4.12) are described, starting with individual component behavior due to strong interactions with other submodels, including central control.

4.4.1 Individual Component Behavior

Manufacturing system elements are modeled as individual components/submodels. This approach is chosen (a) to be able to model individual element behavior in detail in a structured and easily adaptable manner, (b) to provide a modular, fast configurable approach with defined interfaces between modules, represented by different models and (c) to provide the necessary flexibility for representing different system structures and to allow scalability of the concept by creating/destroying and connecting parameterized instances of submodels. Within the following paragraphs, models for manufacturing processes, buffers, and products will be described in detail (compressors are also modeled as single modules, but described within the compressor park section).

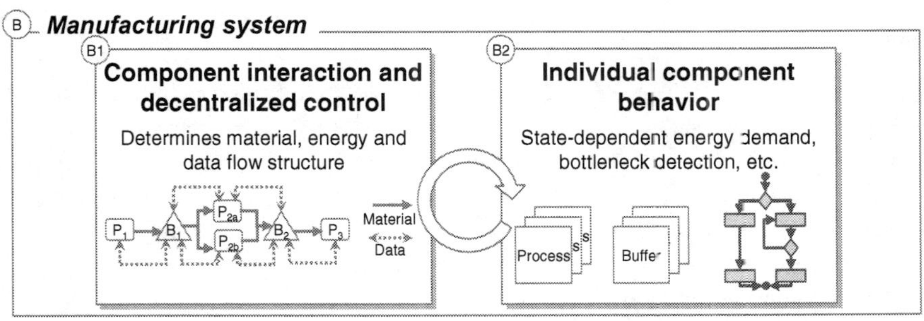

Fig. 4.12 Manufacturing system elements (compare also to Fig. 4.4, own illustration)

Manufacturing Process Model

Connected manufacturing processes and buffers (modeled within the component interaction and decentralized control submodel) constitute the system's material flow logic and information exchange between processes. The total system consists of $N \in E$ processes with index $n = 1, \ldots, N$ and $N + 1$ buffers, while each process has a preceding buffer and the last process has an additional outgoing buffer. Processes are classified into binary ($P_n^{bin.}$) and continuous processes ($P_n^{con.}$). Binary processes can convert a product at a given cycle time Ct_n (e.g., in seconds), which is assumed to be fixed, i.e., production speed cannot be (significantly) adjusted. Examples for binary processes include turning, grinding or milling. This logic assumes that a desired production speed exists which yields targeted results (e.g., product quality, specific energy demand, tool lifetime), although production speed could be adjusted (e.g., adjusting spindle rotation speed/feed rate) (Kara and Li 2011; Yusta et al. 2010). Continuous processes have a minimum cycle time $Ct_n^{con.,min}$, which reflects the shortest possible product conversion time of the process. However, cycle time can be continuously increased (and thus rate decreases, i.e., transformation speed is reduced), examples include conveyor belts or automated transportation systems. The term production rate $R_n = 1/Ct_n$ (e.g., in products/second) is utilized interchangeably with cycle time, for example, to simplify equations (note: $Ct_n^{con.,min} = 1/R_n^{con.,max}$ and $R_n^{con.} \in [0, R_n^{con.,max}]$). Rate and cycle time can change depending on product type for processing. As the proposed concept considers discrete manufacturing systems, products are assumed to be countable. Total number of produced products (cumulative production) of process n until time t is denoted as $CP_{n,t}$ with

$$CP_{n,t} = \int_0^t R_{n,t} dt. \tag{4.31}$$

State-based modeling is used to describe manufacturing process behavior and energy demand, similar to, e.g., Li and Kara (2011), Shrouf et al. (2014), Duflou et al. (2012), for a detailed overview of state-based energy demand modeling see, e.g., Thiede (2012). The corresponding state-chart can be found in Fig. 4.13. Dynamic processing/material conversion rate, resulting electricity demand $El_{n,t}$ (e.g., in watts) and CA demand $CA_{n,t}$ (e.g., in Nm^3/min.) is determined considering the following states:

- **Off**: Process n is turned off ($R_n^{off} = 0$) and requires a fixed electricity amount El_n^{off} and compressed air amount CA_n^{off} (note: for most cases both electricity and compressed air demand can be assumed to be zero when the machine is off). Initial machine state (at beginning of the system evolution calculation) is assumed to be *off*.
- **Switching on**: If a manufacturing process receives a signal to enter into idle or production mode, and the process is currently switched off, the process is required to enter switching on state before entering idle mode (and subsequently production state). Production rate remains zero, electricity and CA demand are $El_n^{turn\ on}$ and $CA_n^{turn\ on}$, respectively. Switching a process on requires a certain amount of time,

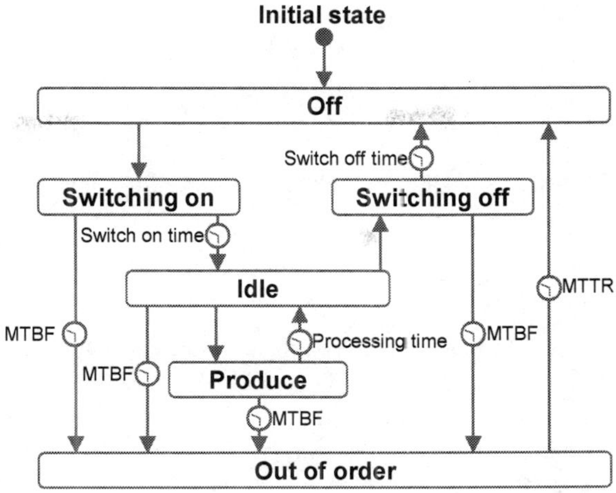

Fig. 4.13 Manufacturing process state-chart (MTBF: mean time between failures, MTTR: mean time to repair, own illustration)

denoted by $SOnT_n$ (switch on time, e.g., in seconds). After passing this given amount of time, the process assumes idle state.

- **Switching off**: Similar to switching on, switching off can only be entered from idle state (i.e., processing of a part needs to be finished before switching off) and is initiated by a control signal to switch the process off. Production rate is zero, further parameters are $El_n^{turn\ off}$ and $CA_n^{turn\ off}$, required time is $SOffT_n$ (switch off time, e.g., in seconds).
- **Idle**: Idle state denotes a machine state during which the machine is not performing work but can (immediately) switch to processing a part. Production rate is zero ($R_n^{idle} = 0$) and the process demands a fixed electricity amount El_n^{idle} and CA amount CA_n^{idle}.
- **Produce**: If an input product is available and the process is scheduled for production, product conversion starts. Process n converts a product at time t with a certain rate $R_{n,t}$ ($R_{n,t} = R_n^{max}$ for binary processes). Energy demand is assumed to scale linear with rate above idle demand, and therefore electricity demand is

$$El_{n,t}^{prod.} = El_n^{idle} + El_n^{spec.} \cdot R_{n,t} \tag{4.32}$$

and CA demand

$$CA_{n,t}^{prod.} = CA_n^{idle} + CA_n^{spec.} \cdot R_{n,t}, \tag{4.33}$$

where $El_n^{spec.}$ (e.g., in watt-seconds per product) and $CA_n^{spec.}$ (e.g., in Nm^3 per product) denote the rate-dependent electricity and CA demand of a process. Consequently, a process' current variable electricity demand is $El_{n,t}^{var.} = El_n^{spec.} \cdot R_{n,t}$, variable CA demand is correspondingly determined. Product conversion ends when a full machine cycle has been passed. For binary processes, a conversion cycle that starts at time $t_{n,start}$ ends at time $t_{n,end} = t_{n,start} + Ct_n$. For continuous processes and changing conversion rates during production, conversion ends according to

$$t_{n,end} = t_{n,start} + \Delta t_{n,conv.} \quad (4.34)$$

with $\Delta t_{n,conv.}$ being the required time for conversion, which is determined by

$$\int_0^{\Delta t_{conv.}} R_{n,t}\, dt = 1. \quad (4.35)$$

In other words, conversion is complete when the integral of current (less than maximum intensive) rate over time equals one (product). For example, if a process maximum conversion rate is four products per time unit, i.e., one product every quarter time unit, and the process operates at two products per time unit (half speed), one conversion cycle (one product) is completed after half of a time unit. For an illustrative example see Fig. 4.14. Both example processes from the figure start with their maximum conversion rate. Process two then switches to a lower rate for a given amount of time ($R_2 < R_2^{max}$, i.e., to reduce energy demand), after which the process resumes speed to its maximum rate. The reduced operating speed period delays completion of process two compared to process one ($t_{1,end} < t_{2,end}$). After product conversion has been finished, the process enters idle state and another product conversion cycle can be started, the process can be switched off or simply remain in its idle state.

- **Out of order** Aside from its off-state, the process can enter into out of order state from any other state. Transitions are randomly triggered, probability is characterized by mean time between failures (MTBF, for detailed description on failure behavior see below). If out of order is entered from produce state, product conversion is interrupted and the currently processed part removed from the process and disposed of (assumes that machine failure results in damaged/rejected parts). Repairing the machine requires a randomly distributed mean time to repair (MTTR, see also below for details), after which the machine enters into off-state and can be switched on again.

Empirical studies on the differentiation of energy demand into a fixed (idle) energy demand and rate-dependent demand can, for example, be found in Gutowski et al. (2006) for several processes. For the case of continuous processes, adjustable rate and thus adjustable rate-dependent energy demand results in nonlinear, load dependent energy demand per product if fixed demand is not zero: a conveyor belt which speed is reduced from 20 cm/min to 15 cm/min, with products placed at 5 cm distance reduces its rate from four products per minute to three products per minute. Assuming that fixed energy demand per minute is 30 Wh and rate-dependent demand 40 Wh/min at full speed, energy demand at four products/min would be 70 Wh/min and thus 17.5 Wh/product, while demand at three products/min would be 60 Wh/min and thus 20 Wh per product. For an example of rate-dependent energy demand, especially of conveyor belts, see, e.g., Deutsches Institut für Normung e. V. (2011).

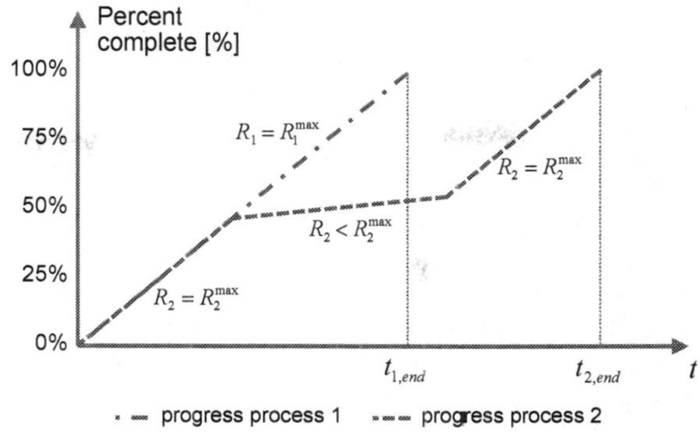

Fig. 4.14 Example for processing speed with varying cycle time and constant (maximum) cycle time (own illustration)

Further, for a multiproduct case and $ty = 1, \ldots, TY$ denoting the total number of different product types, variable energy demand of processes is changed depending on product type currently manufactured on the machine.

Failure behavior of machines is assumed to underlie random behavior, which includes both failure and repair times. MTBF is sometimes referred to as *mean operating time between failures* or *mean time between failures* (Birolini 1999). In contrast to *mean time to failure* (MTTF), which describes failure behavior of non-repairable machines, MTBF is used here to describe failure behavior of repairable machines, i.e., machines can be repaired and are considered equivalent to a new machine after repair with respect to their failure behavior (i.e., MTBF is independent and identically distributed (i.i.d.), Bertsche and Lechner 2004). Availability Av_n of process n is

$$Av_n = \frac{MTBF_n}{MTBF_n + MTTR_n} \tag{4.36}$$

and denotes the average (expected) availability of a process as a share of total time, i.e., what percent of total time is the process expected to be available for normal operation (Birolini 1999).

Random failure behavior can be described by various probability functions, depending on the nature of failure behavior (and thus statistical fit of a chosen function). Due to its flexibility and wide universal applicability, the two-parameter Weibull distribution is widely used to statistically describe process reliability (Birolini 1999; Dodson and Nolan 2002; Bertsche and Lechner 2004; Wilker 2010; Thiede 2012). Its probability density function is

$$f(t) = \frac{\beta}{\alpha} \cdot \left(\frac{t}{\alpha}\right)^{\beta-1} e^{(-t/\alpha)^{\beta}} \tag{4.37}$$

with β being the shape parameter of the function and α the scale parameter, also known as characteristic lifetime. For the purpose of describing failure behavior of machines, the case $\beta = 1$ denotes an independent failure behavior, i.e., constant failure rate (as opposed to a decreasing failure rate, e.g., at the beginning of a components life and increasing failure rate at its end). Equation 4.37 is then equivalent to an exponential distribution. For estimating statistical distribution parameters for MTBF from observed reliability data, see, e.g., Wilker (2010) or Dodson and Nolan (2002). MTTR can be modeled using, e.g., a uniform, log-normal or Erlang distribution, depending on statistical fit (Chung 2004; Sterman 2000; Schultz et al. 2015).

Processes have an internal logic implemented to avoid throughput loss and to provide an interface to the central production and energy flexibility control. Two more process states are introduced, coined by the term *criticality*. Critical state is assumed when the process impacts or is likely to impact overall system throughput. A process P_n can enter a critical state due to the following reasons:

- The process P_n is the current bottleneck of the system.
- The process P_n is likely to directly or indirectly block the current bottleneck of the system, i.e., the upstream buffer reaches a critical high fill level and the upstream process P_{n-1} is critical (upstream process is blocked when the process cannot deliver finished products to its outgoing buffer).
- The process P_n is likely to directly or indirectly starve the current bottleneck of the system, i.e., the downstream buffer reaches a critical low fill level and the downstream process P_{n+1} is critical (downstream process is starved if no input products are available in its incoming buffer).

Only the first condition is not dependent upon states of upstream or downstream processes. Consequently, only the bottleneck process(es) can enter critical state first and then cause criticality at other processes, dependent on intermediate buffer fill levels. If a process is critical needs to be newly determined every time influencing conditions change (e.g., process rate, connected buffer fill levels, up-/downstream process states, etc., cf. Fig. 4.15). As the bottleneck of the system can only be determined by considering other system elements, bottleneck detection will be described in Sect. 4.4.2 *Component Interaction and decentralized control.*

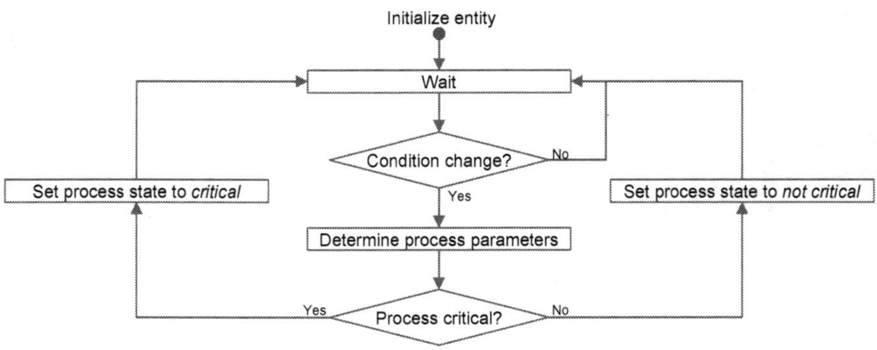

Fig. 4.15 Flowchart for determining process criticality (own illustration)

In summary, modeled process behavior is generic and can reflect a variety of different manufacturing and production processes. Through adequate parameterization, individual processing and energy demand behavior can be reflected. The second main manufacturing system element are product buffers, which are described in the next section.

Buffer Model

While converting a product, a single product is stored within a given process. All other products in the system are required to be in a buffer. Consequently, all product storage spaces, including multiple products in automated transport systems (e.g., conveyors) need to be modeled as buffers with limited capacity. Buffer content (fill level $FL_{n,t}$) at time t is determined by

$$FL_{n,t} = \max\left(0, FL_{n,0} + \sum_{a=c}^{d} CP_{a,t} - \left(\sum_{k=q}^{r} CP_{k,t} + (r-q)\right)\right) \tag{4.38}$$

where $c, \ldots, d$ denotes all processes which feed products into the buffer (all connected upstream processes) and $q, \ldots, r$ all processes which draw products from the buffer (all connected downstream processes). Buffer fill level according to Eq. 4.38 is determined by adding cumulative production from upstream processes since model start and subtracting cumulative production since model start and subtracting $r - q$ products, which are currently stored within the processes for processing, if parts were already available for processing since the start of the model (max-function for the case that all input values are zero).

Product Model

Products are modeled as submodels to allow tracking of an individual product when it passes through the system and to assign characteristic values (e.g., current value, embodied energy, energy demand depending on process type). In general, products can have certain characteristics that are used as input for other system elements (e.g., different product types can be defined to determine processing (cycle) times and variable energy demand of manufacturing processes), or (changeable) characteristics which describe the current state of the product.

Product characteristics used for input into other models are, for example, product type-specific variable electricity and CA demand, as well as cycle time per process. Upon arrival of a product at a process, the product obtains its current processing state, i.e., at which process (type) the product has arrived. After acquiring the current process type, the product determines, according to predefined values (e.g., using a product–process type table for energy demand and other parameters), what parameters to communicate to the current process. The process then adjusts its parameters (e.g., energy demand) according to values submitted by the current product.

Changeable characteristics of products are used to attach product-specific values to products, where $l = 1, \ldots, L$ denotes the index for a specific product, with total number of products currently in the system being L. These values can be of any product-specific kind. Proposed concept includes the product-specific monetary value PrV_l (e.g., in EUR), entry time into the system PrT_l (e.g., minutes) and embodied grid, wind and solar electricity PrE_l^{grid}, PrE_l^{wind}, PrE_l^{solar}.

Monetary value of a specific product is calculated by adding processing costs to the product's initial value. Assuming that product value before entering a process n with processing start time t^{start} and processing end time t^{end} is $PrV_l^{init.}$ and the given product just finished processing at n, total product value is

$$PrV_{l,t^{end}} = PrV_{l,t^{start}} + \Delta PrV_{l,n} \tag{4.39}$$

with $\Delta PrV_{l,n}$ denoting the monetary value change due to processing product l on process n. $\Delta PrV_{l,n}$ is calculated utilizing a process-specific hourly rate PHR_n (e.g., in Euro per hour, here: per hour, similar to a machine hourly rate used, e.g., in accounting) and product-specific minimum cycle time $Ct_{n,l}$ (here in seconds):

$$\Delta PrV_{l,n} = Ct_{n,l}^{min} \cdot PHR_n \cdot 1/3600. \tag{4.40}$$

For simplicity, minimum cycle time is utilized rather than real processing time depending on rate increase or decrease of continuous processes, as this would include other simplifying assumptions such as constant cost independent of processing speed (including energy cost).

Embodied electricity values are required in order to track and remove differences from initial and final system states to enable evaluation of system performance independent of the system's start and end states. Further, tracking embodied energy enables determining what energy mix has been used to process a given single product, which can be used to increase accuracy of further (environmental) impact estimation methods. Embodied electricity is calculated by adding source-specific electricity demand during processing to the specific product's embodied electricity. Embodied electricity $PrE_{l,s,t^{end}}$ (in Wh) of product l from source s after processing at machine n with processing start time t^{start} and processing end time t^{end} is calculated by adding direct electricity demand from processing the part, embodied electricity in compressed air and electricity stored in a battery and/or EV(s) to its initial embodied electricity on a per source basis:

$$\begin{aligned} PrE_{l,s,t^{end}} = PrE_{l,s,t^{start}} + \int_{t^{start}}^{t^{end}} [\phi_{s,t}^{supply} \cdot El_{n,t}^{direct} \\ + \phi_{s,t}^{CCA} \cdot EICA_t \cdot CA_{n,t} + \phi_{s,t}^{bat.} \cdot El_{n,t}^{bat.}]\, dt, \end{aligned} \tag{4.41}$$

with $El_{n,t}^{direct} + El_{n,t}^{bat.} = El_{n,t}$ (split of electricity demand into demand satisfied from immediate generation and from battery), $\phi_{s,t}$ denoting the current share of a given supply source from total energy flow supplied by this source and $EICA_t$ the time-dependent electricity intensity of compressed air (e.g., in Wh/Nm^3) (cf. Sect. 4.3.5). As mentioned before, source supply share in CA is dynamically tracked, also re-

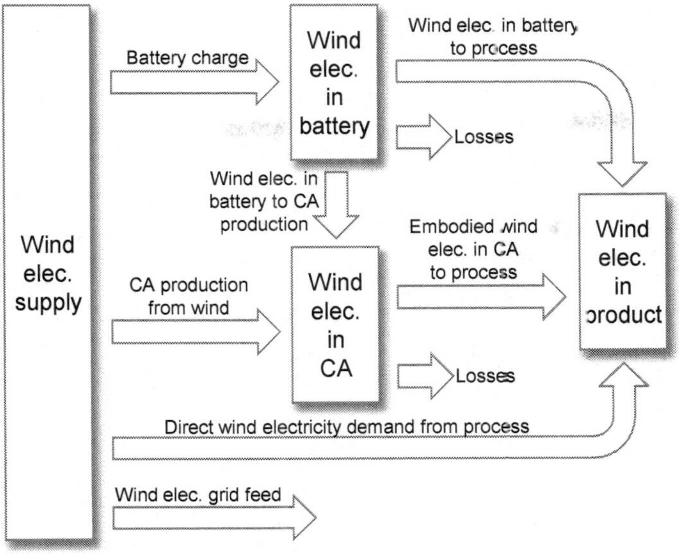

Fig. 4.16 Flow chart for embodied wind electricity in a product (own illustration)

flecting that electricity from battery can be supplied to compressors, which is source specific and thus needs to be reflected in Eqs. 4.27–4.30 (not further detailed to limit complexity). The complete flow chart for determining embodied electricity for a product can be found in Fig. 4.16, utilizing wind as an example supply source. Total energy supply share of a given electricity source varies over time. In combination with both battery and compressed air energy storage, total source-specific energy embodied in a product due to processing at a given machine is highly dynamic.

Embodied electricity in products, compressed air, and battery energy storage needs to be tracked to compare different system configurations. For example, if grid feed-in of RE is monetarily compensated, the difference in embodied RE electricity and its value currently contained in the system needs to be taken into consideration when economically comparing different scenarios. For example, scenario one results in high product inventory levels and high embodied RE electricity in inventory at the end of system evolution calculation. Scenario two ends with low inventory levels and a higher RE feed-in. Consequently, scenario two would receive a higher monetary compensation for RE feed-in, which needs to be removed by adding the value of embodied RE electricity in scenario one. This method is essential to ensure independence from system start and end states (and overall model time and evolution calculation). This logic is true for all indicators which are affected by start and end states, e.g., energy self-sufficiency ratios and emission calculation (e.g., CO_2 intensity, calculated by applying a supply source-specific emission factor, detailed in the following description of indicators).

4.4.2 Component Interaction and Decentralized Control

The second main element of the manufacturing system model (B1 in Fig. 4.12) is the component interaction and decentralized control. Previously described individual elements of the manufacturing system and auxiliary system(s) realize a structured and dynamic material and energy flow through the system via their interaction logic and structure. This section focuses on the interaction of all system elements, centered around the manufacturing system and its decentralized control strategy. For an overview of manufacturing system energy flows see, e.g., Posselt (2016).

The manufacturing system, in particular the manufacturing process structure, establishes the material flow logic through the system. Figure 4.17 shows an example manufacturing system process structure, which will be utilized as illustrative example throughout this section.

A process chain within this concept exhibits the following characteristics:

- The process chain always starts with an incoming and ends with an outgoing buffer, which are never empty or full (assumes that sufficient raw material and final product space is available at all times). This logic creates a material flow system boundary, which does not exogenously affect overall system behavior (i.e., insufficient raw material would starve the first process and thus affect system behavior).
- A process is always connected to an incoming buffer for product input and an outgoing buffer to deliver processed products (also indirectly via dividing and combining flows). Buffer size can be equal to one product, denoting that products are directly transferred to another process for immediate processing.
- Dividing flow elements (FD) are connecting one buffer with two processes. A flexible allocation logic which process receives a part can be implemented if only one part is available and both processes are waiting for a part. Within the proposed concept, the initial logic for dividing product flows is dependent upon input requirement of the two downstream processes. This requirement is based on blockage and starvation times of respective processes and will be further detailed within the central electricity control logic section.

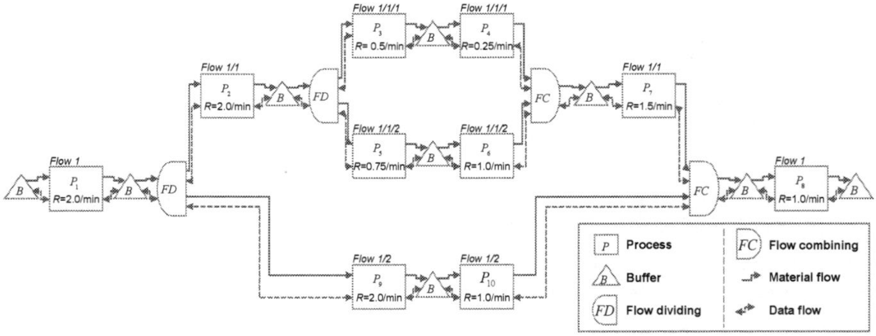

Fig. 4.17 Example manufacturing system process structure (*R*: production rate, own illustration)

- Combining flow elements (FC) always connect two processes with one buffer. If the jointly utilized buffer is full, a logic needs to be implemented which process is allowed to deliver a final product. For the purpose of this concept, a first-come-first-served logic is applied, i.e., the process which finished processing first is allowed to deliver a product into the buffer if remaining buffer space is only one item.
- Material flow is unidirectional, i.e., no reverse flows are considered.
- All manufacturing system elements are exchanging data through buffers and FD/FC elements to, e.g., determine available outgoing buffer space or for bottleneck detection purposes.

A central role for decentralized control of manufacturing processes is bottleneck detection of the system. Several methods have been proposed for (dynamic) bottleneck detection (for an overview see, e.g., Li et al. 2007, Zhai et al. 2011, Roser et al. 2014). No uniform definition for what a bottleneck constitutes can be found in literature (for a list of definitions see, e.g., Roser et al. 2014). Within this work and based on existing literature, a bottleneck shall be defined as the process that currently limits throughput of the manufacturing system, i.e., the system bottleneck might change depending on system state changes.

Different methods are known to detect a serial manufacturing line bottleneck, among them utilizing simulation or applying an analytic method (e.g., Chiang et al. 1998). Detecting a bottleneck with analytical methods is only feasible for production lines with two processes as no closed form expression for system production can be found for systems with more than two processes. Otherwise, other methods such as heuristics (e.g., Mason et al. 2002) or approaches involving simulation (e.g., Mönch and Zimmermann 2007) need to be applied.

Further, a method considering manufacturing blockage and starvation frequencies has been proposed in Kuo et al. (1996), which can be applied using either calculated or measured (online) data. Dynamic bottleneck detection utilizing a so-called 'bottleneck walk' has been proposed in Roser et al. (2014). A data-driven method for bottleneck detection is suggested in Li et al. (2009), which has been applied to a manufacturing line in an energy flexibility context in Zhou and Li (2013).

Within Sect. 4.4.1, the term *process criticality* has been introduced to characterize processes which are throughput critical. In order to determine the bottleneck process and bottleneck-influencing processes, any methodology can be applied to identify the bottleneck process. Within this concept, the following method is applied for dynamic bottleneck identification: the system bottleneck is the process (or processes) which are expected to yield the lowest cumulative production over total model time of all processes, or, for the case of parallel flows, the processes that are jointly expected to yield the lowest cumulative production. In particular, a process P_{bn} with maximum production rate R_{bn}^{max} and expected availability Av_{bn} is bottleneck of a serial production line with no diverting or combining flows at time t if

$$CP_{bn,t} + R_{bn}^{max} \cdot (T - t) \cdot Av_{bn} \leq CP_{n,t} + R_{n}^{max} \cdot (T - t) \cdot Av_{n} \quad \forall n \neq bn. \tag{4.42}$$

In other words, the process with the lowest sum of already processed products plus expected maximum production during remaining model time, considering expected availability, is the bottleneck processes.

For diverting and combining flow structures, bottleneck detection applying above method requires consideration of local (parallel flow) bottlenecks. Maximum total expected throughput of a parallel flow needs to be determined (combined throughput potential) and compared to throughput values from processes/other parallel flows which are not part of the considered parallel flow. Utilizing the example structure from Fig. 4.17, three process structure levels can be identified: all products need to pass through level one (labeled *Flow 1*, i.e., P_1), while products can either pass through the first subflow of level two (denoted as *Flow 1/1*, e.g., P_2) or its parallel flow (*Flow 1/2*, e.g., P_9). Within the first subflow of level two, the products can enter into two additional parallel flows which are part of process structure level three (*Flow 1/1/1*, e.g., P_3 and *Flow 1/1/2*, e.g., P_5). In order to determine the bottleneck of the overall system, the bottleneck of each parallel flow is determined first, starting with the lowest level. Assuming $t = 0$ and thus $CP_n = 0$ for all processes, and a total model time $T = 100$ min and no failures ($Av_n = 1$), P_4 is the bottleneck of subflow 1/1/1 with expected production of 25 parts. P_5 is the bottleneck of flow 1/1/2 with expected production of 75 parts. Thus, both flows combined have an expected output of 100 parts, which is used as input to determine maximum throughput of flow 1/1. Considering the rates from P_2 and P_7, which are also part of flow 1/1, the parallel flows 1/1/1 and 1/1/2 and their processes P_4 and P_5 determine maximum throughput of flow 1/1 as expected throughput of P_2 and P_7 is higher than the combined 100 pieces from 1/1/1 and 1/1/2 (namely 200 (P_2) and 150 (P_7) pieces). Therefore, maximum remaining throughput capacity of flow 1/1 is 100 pieces. P_{10} is the bottleneck of flow 1/2 (100 pieces remaining capacity), and consequently, combined throughput of flow 1/1 and 1/2 is 200 pieces (with bottleneck processes P_4, P_5 and P_{10}). Comparing this value with all processes which are only part of level one (P_1 and P_8), P_8 has the lowest remaining capacity with 100 pieces and therefore determines total system throughput (and consequently is the single bottleneck process). However, if cumulative production values change when model time advances, bottlenecks might shift depending on process utilization rates. The current bottleneck is dynamically determined using described method. Processes exchange data via buffers to determine their relative position in the total system structure (i.e., what sublevel(s) they are part of). Flow dividing and flow combining elements determine the combined maximum throughput of parallel flows. Further, in order to let processes know that they are a bottleneck process, total system throughput is first evaluated according to described method, and then FD and FC elements detect if a parallel flow structure is part of the bottleneck. If a parallel structure is part of the bottleneck, parallel flows determine which processes are limiting throughput of each parallel flow and are thus part of the bottleneck (if the parallel flows are part of the system bottleneck).

System throughput plays a central role for evaluating and comparing different production management and thus energy flexibility strategies. If overall throughput varies between different model evolution calculations, the differential needs to be reflected in indicators such as energy demand, throughput time or economic evaluations. For example, reduced throughput results in lost contribution margin of sales and therefore inherently increases costs for load shifting actions (see, e.g., Paulus and Borggrefe 2011). In addition, lost/reduced throughput may result in less acceptance of load shifting actions on the side of decision makers (Sun and Li 2014). Consequently, to eliminate a free variable, and assuming that a target production exists (i.e., due to delivery requirements), proposed concept aims at realizing constant throughput. Constant throughput is realized by introducing a virtual customer at the end of the process chain. For the example case in Fig. 4.17, P_8 would be equivalent to a virtual customer process. The customer process is excluded from any energy control actions, is always 'producing', demands no energy and has a fixed cycle time $Ct_{cust.}$. This cycle time denotes the total production requirements, for example, a cycle time of one minute automatically results in a production of 480 pieces in an eight-hour shift. Customer cycle time can either be arbitrarily chosen according to the initial capacity of the manufacturing line to achieve a maximum throughput or according to (external) delivery requirements or target production rate. Note that the customer process needs to be initially the bottleneck, otherwise total production targets cannot be achieved as throughput is determined by another process.

For a multiproduct case, achieving constant throughput via bottleneck detection increases complexity. However, if a mandatory production program is known, and processing times do not vary depending on which route a given product type takes in parallel flow structures, bottleneck detection can be enacted similar to above example. Cumulative maximum production needs to be evaluated per product type and bottlenecks are detected for each product type. However, if processing times vary depending on product routing, above bottleneck detection cannot be applied unless chosen route is ex-ante known. As a consequence, bottleneck detection and thus throughput control is not enacted for multiproduct scenarios and throughput needs to be compared between different scenario evaluations.

4.5 Central Energy Control

The third main element of proposed real-time system behavior and control model is the central energy control model (Fig. 4.18). Its main purpose is to enable energy flexibility via controlling processes' energy demand (electricity and compressed air) in relation to available (on-site generated VRE) electricity. As noted in the theoretical background on energy flexibility (Sect. 2.1.4), enabling or increasing energy flexibility of a manufacturing system is closely related to embodied energy storage in products. Increasing the (electricity) supply share of VRE during processing a product implicitly "stores" VRE in products (embodied energy). As such, intermediate product storage becomes also embodied energy storage. Increasing the amount of (intermediate) products within a product buffer therefore increases the amount of embodied energy which is stored. Energy flexibility can be used to control the type of energy that is stored within a product, for example, by controlling processes' energy (e.g., electricity and CA) demand. This includes indirect (auxiliary systems)

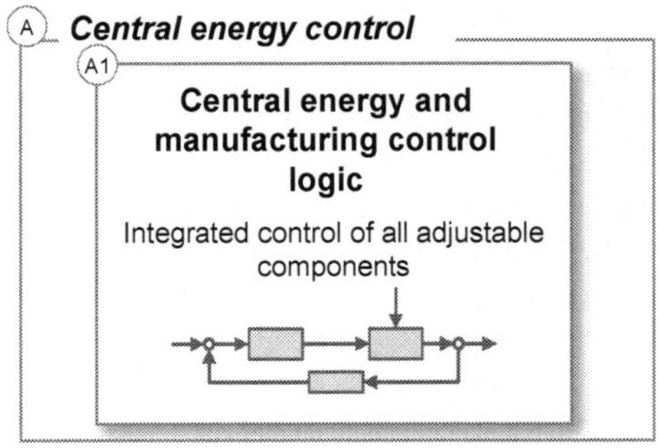

Fig. 4.18 Central energy control element (compare also to Fig. 4.4, own illustration)

energy demand, for example, electricity demand of compressors, which is transformed through manufacturing CA demand into embodied energy in a product. However, in order to increase the share of VRE in embodied energy in products, not only energy flexibility of a process or system is required, but also a control logic to use energy flexibility of a system to increase VRE demand. This control logic should be integrated into the wider manufacturing system control to account for dependencies, especially with manufacturing operations and material flow. This sections outlines an execution-related approach to an integrated energy control to influence energy demand of a manufacturing system towards increased VRE utilization.

As outlined in Sect. 3.6, existing research mainly focuses on optimization/planning methods. Consequently, proposed concept aims at providing a real-time control method, which can be implemented within a manufacturing execution control. The main reason for a real-time approach is to account for stochastic behavior of system elements. This includes, for example, variable (real-time) energy supply, i.e., forecasting is not mandatory and forecasting errors do not result in a deviation from a given (predetermined) plan. Further, stochastic behavior of any system element (e.g., machine and compressor energy demand due to failure) can be accommodated as the control considers the current system state as input (i.e., using online data), output signals are obtained based on observed data and thus energy demand is adjusted according to current overall system state.

The central energy control logic's main goal is to maximize demand of on-site generated electricity. Figure 4.19 gives an overview of the closed loop control structure described in this section (for an overview and control system theory see, e.g., Aström and Murray 2008). The reference value/input is the currently available electricity which is generated on-site (El_t^{own}). This value is compared to current total electricity demand of the system ($El_{sys.,t}^{dem.}$). Based on obtained delta $\Delta El_{sys.,t}^{dem.} = El_{sys.,t}^{dem.} - El_t^{own}$, the central electricity control determines required control actions (note that $\Delta El_{sys.,t}^{dem.}$ is equivalent to grid supply or feed-in, El_t^{grid}). Total system electricity demand cannot be directly controlled (i.e., via a single control interface or switch), therefore electricity demanding and controllable system elements (e.g., manufacturing processes, compressors) receive target state control signals which are expected to yield the closest fit between electricity demand and on-site supply. Depending on current system state,

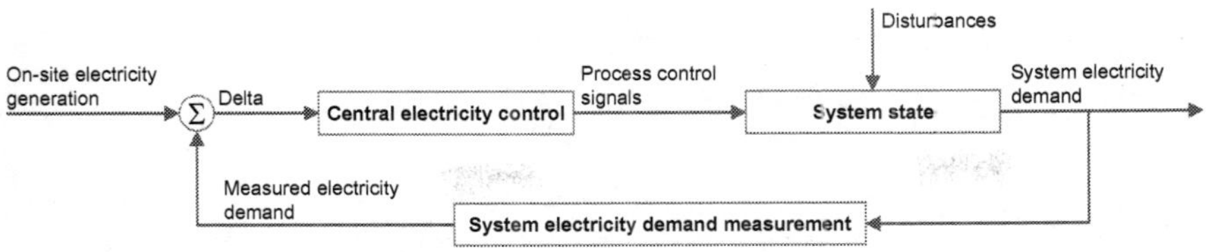

Fig. 4.19 Central electricity control loop (own illustration)

states of individual elements and calculated control signals, the system changes into a new state. External disturbances, for example, process failures, product processing changes (e.g., finished processing) or process criticality changes (throughput critical or CA pressure criticality) influence the system state. Consequently, a constantly changing system determines the system electricity demand, which is then measured and fed back as new input into the control loop.

Within the remainder of this section, the control logic is described in detail. A so-called *control cycle* is executed to match electricity demand and supply. A control cycle is triggered if at least one of the following conditions is true:

- Own electricity supply changes.
- System electricity demand changes due to process self-induced state changes (e.g., due to finished processing, out of order).
- The set of adjustable processes changes, i.e., due to criticality changes or failure.

The central electricity control logic applies two different control methods to achieve a good fit with available electricity supply. First, binary processes, which offer a limited degree of flexibility due to their discrete electricity demand levels, are rescheduled. Depending on obtained fit with available supply, continuous processes are adjusted to improve overall system demand fit (compare "adjustability" also to energy flexibility evaluation of processes in, e.g., Graßl 2015 and Graßl et al. 2013).

In particular, all adjustable binary processes $ab = 1, \ldots, AB$ with $AB \in E$ are determined. A process is adjustable if not critical due to throughput or CA pressure requirements (compressor), if not out of order or mandatory switched off (e.g., due to energy saving purposes).

In a first step, target electricity demand for adjustable binary processes is determined by

$$TEL_t^{bin.} = \max\left(0,\ El_t^{own} - \left(El_{sys.,t}^{sched.} - El_t^{bin.,var.}\right)\right), \tag{4.43}$$

with $El_{sys.,t}^{sched.}$ being the currently scheduled electricity demand of the system, i.e., the electricity demand of the system if all processes would produce according to their scheduled production rate. Resulting/observed electricity demand $El_{sys.,t}^{dem.}$ can be lower than scheduled system demand if processes are blocked or starved. Further, $El_t^{bin.,var.}$ denotes the current variable electricity demand of all binary processes which are adjustable, with

$$El_t^{bin.,var} = \sum_{ab=1}^{AB} El_{ab}^{spec.} \cdot R_{ab,t}. \tag{4.44}$$

Equation 4.43 therefore determines the target electricity demand for adjustable binary processes as the total available own electricity minus nonadjustable electricity demand (term in inner parenthesis). Nonadjustable electricity demand is all (scheduled) electricity demand except adjustable demand from binary processes, i.e., nonadjustable demand includes process idle demand, demand from continuous processes, and demand from nonadjustable binary processes.

After target electricity has been calculated, process control signals need to be specified. Controllable binary process states are *idle* and *produce*, as switching a process on or off would require a given amount of time. Within this concept, changes in own supply are assumed to occur very frequently (i.e., minute-based changes), which requires a (nearly) immediate change in electricity demand. As a result, time constants for on/off switches are omitted in a first step within the central electricity control (note that decentralized process self-switch-off is used to switch processes off according to current process state and adjacent buffer inventories to reduce (external) electricity demand, discussed later). To control electricity demand according to available electricity, all possible idle/produce combinations for all adjustable binary process AB are determined. Total number of combinations is $NCP = 2^{AB}$, where $ncp = 1, \ldots, NCP$ is the index for a specific control signal combination. Corresponding variable electricity demand of adjustable processes in a given combination is then calculated by

$$El_{ncp} = \sum_{ab=1}^{AB} El_{ab}^{spec.} \cdot R_{ab}^{max} \cdot \nu_{ab} \tag{4.45}$$

with $\nu_{ab} = \{0, 1\}$ denoting the control signal (idle/produce) for process ab in combination ncp.

After variable electricity demand of all combinations has been calculated, the combination ncp_o with the closest fit to target electricity demand is evaluated:

$$|El_{ncp_o} - TEL_t^{bin.}| \leq |El_{ncp} - TEL_t^{bin.}| \;\; \forall \; ncp \neq ncp_o. \tag{4.46}$$

Adjustable binary processes then receive control signals according to the optimal control combination ncp_o. However, as processes might not be able to assume idling immediately (i.e., product conversion needs to finish first) or the optimal combination yields no perfect fit with target electricity demand ($El_{ncp_o} \neq TEL_t^{bin.}$), a gap between actual electricity demand and on-site generated electricity might still exist.

In order to close this gap, continuous processes are scheduled in a second step. This method is chosen as continuous processes offer a higher flexibility due to their linear production rate adjustability and thus continuous energy demand changes.

First, required electricity adjustment is calculated similar to above, but calculation takes immediately place after binary process adjustment has been performed: $\Delta El^{dem.}_{sys.,t} = El^{dem.}_{sys.,t} - El^{own}_{t}$. For each adjustable (noncritical, not out of order, not mandatory switched off) continuous process $ac = 1, \ldots, AC$ with $AC \in E$, a maximum electricity adjustment potential $ElAP_{ac}$ (e.g., in watt) is calculated by

$$ElAP_{ac,t} = \begin{cases} El^{spec.}_{ac} \cdot \left(R^{max}_{ac} - R_{ac,t}\right) & \text{if } \Delta El^{dem.}_{sys.,t} < 0 \\ El^{spec.}_{ac} \cdot \left(R_{ac,t} - R^{min}_{ac}\right) & \text{if } \Delta El^{dem.}_{sys.,t} > 0. \end{cases} \tag{4.47}$$

Above equation determines the maximum potential increase of electricity demand from adjustable continuous processes if system electricity demand is lower than own supply by calculating the change in demand if all processes were set to their maximum rate. If system demand is higher than own supply, the adjustment potential is based on the change to minimum rate for all processes.

If maximum adjustment potential of processes is lower than required adjustment,

$$\sum_{ac=1}^{AC} ElAP_{ac,t} \leq |\Delta El^{dem.}_{sys.,t}|, \tag{4.48}$$

all processes receive a signal to set their production rate to their minimum (own supply lower than demand) or maximum (own supply higher than demand) rate.

If the sum of all adjustment potentials is lower than the absolute adjustment requirement, a logic for determining individual processes rate adjustment needs to be applied. As noted before, proposed concept aims at preventing throughput loss due to load management actions. Therefore, a buffer-based logic is applied to determine allocated rate adjustment. If electricity demand needs to be reduced, process rate is adjusted depending on a process' time window TW_{ac}, which is the lower value until a process starves outgoing processes or blocks incoming processes if set to a zero production rate. Therefore, a manufacturing process' n (note that indexing refers to all processes within the system) time window is

$$\begin{aligned} TW_{n,t} &= min(BT_{n,t}, ST_{n,t}) \\ \text{where } BT_{n,t} &= \frac{C_n - FL_{n,t}}{\sum_{a=c}^{d} R^{max}_{a}} \\ \text{and } ST_{n,t} &= \frac{FL_{n+1,t}}{\sum_{k=q}^{r} R^{max}_{k}} \end{aligned} \tag{4.49}$$

where $BT_{n,t}$ is the block time of process n, i.e., the time until the incoming buffer would be full if incoming processes $c, \ldots, d$ would produce at their maximum rate while process n does not produce. $ST_{n,t}$ is the starve time until process' n outgoing buffer is empty if all following processes $q, \ldots, r$ would produce at their maximum rate (and n is set to idle/production rate is zero).

In order to obtain a relative measure to allocate reduction/increase requirements for all adjustable continuous processes, time window share $TWS_{ac,t}$ is calculated by

$$TWS_{ac,t} = \frac{TW_{ac,t}}{\sum_{ac=1}^{AC} TW_{ac,t}}. \tag{4.50}$$

Required electricity demand reduction (if $\Delta El_{sys.,t} > 0$) is then allocated to all adjustable processes according to individual time window share, which results in individual reduction requirements $\Delta El_{ac,t}$:

$$\Delta El_{ac,t} = TWS_{ac,t} \cdot -\Delta El_{sys.,t}^{dem.}. \tag{4.51}$$

If individual reduction requirement exceeds a process' maximum adjustment potential ($-\Delta El_{ac,t} > ElAP_{ac,t}$), remaining adjustment requirement is allocated to remaining adjustable processes according to their time window share, which is recalculated without the process that has exceeded its maximum potential (an thus rate is set to the process' minimum rate).

For the case of a demand increase requirement ($\Delta El_{sys.,t} < 0$), processes are also ranked according to their remaining starvation and blocking times (i.e., which processes requires an increase most to avoid production losses/criticality). For this purpose, a priority value $PV_{ac,t}$ is used to rank processes:

$$PV_{ac,t} = \begin{cases} 1/TW_{ac,t} & \text{if } TW_{ac,t} \geq 0.0001 \\ 10{,}000 & \text{if } TW_{ac,t} < 0.0001. \end{cases} \tag{4.52}$$

The function is capped at a value of 10,000 to avoid calculations with variables that go to infinity. The remaining allocation procedure is similar to the allocation algorithm for demand reductions (priority value share is used instead of time window share). For both demand reductions and increases, required electricity demand adjustments are translated into process rate adjustment requirements and thus targets. Corresponding control signals are sent to processes.

Figure 4.20 provides an example for how the central energy control matches electricity demand with own supply. In general, the central electricity control aims at finding the combination of binary processes (produce/idle) that matches on-site supply best, while the remaining adjustment gap (if any) is closed by continuous processes. Process control and thus energy demand are subject to several constraints, e.g., maintaining throughput or maintenance requirements. After considering these constraints, the central electricity control aims at matching supply and demand. If a perfect match is not feasible, surplus electricity is fed into the grid, or, if the system requires electricity, supplied by the grid. Every time the system state changes (changed electricity supply or demand, e.g., if processing for a given process finishes or a machine failure occurs), a new control situation and thus control cycle is triggered.

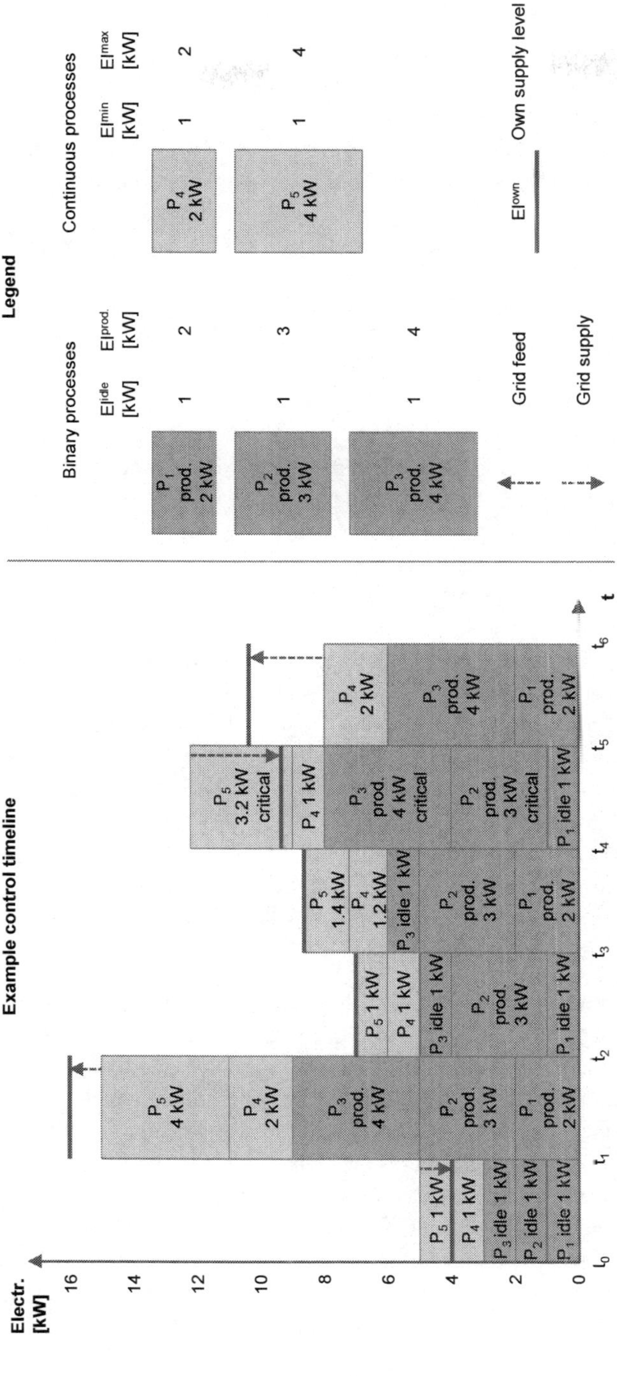

Fig. 4.20 Central electricity control example (own illustration)

Six example control situations (Fig. 4.20) were chosen to illustrate the algorithm, each situation is labeled according to its starting point $t_0, \ldots, t_5$:

- Control situation t_0: Own electricity supply is lower than minimum demand from all processes, thus all processes are set to their lowest electricity demand. Nonetheless, 1 kW needs to be drawn from the grid.
- Control situation t_1: Own electricity supply is higher than maximum (system) demand from all processes. All processes are set to their maximum rate, remaining own supply is fed into the grid.
- Control situation t_2: Available own supply (7 kW) is exactly matched through binary process control: all processes are set to idle (i.e., coming from a state equivalent to t_0, base-load is 5 kW), except process two, which demands an additional 2 kW during production (note that only process two is running and producing, all other processes are in idle mode). This results in an exact match of demand and own supply, which would not have been achieved by any other combination of binary processes' idle/produce states.
- Control situation t_3: This situation demonstrates the effect of continuous process control: if all binary processes would have been switched to production, own supply would have been exceeded, even with lowest production rate from continuous processes. However, continuous control flexibility from processes four and five achieves a fit with own supply by scheduling binary process three to idle and filling the remaining gap by adjusting continuous processes.
- Control situation t_4: Process two, three, and five are critical, for example, due to CA pressure requirements (process two could be a compressor) and throughput requirements (process five could be the manufacturing system bottleneck and process three its input process while low buffer inventories between process three and five are prevailing). Demand exceeds own supply, even if all noncritical processes are set to their lowest rate. Supply from grid is unavoidable.
- Control situation t_5: Process two and five are out of order and thus own supply exceeds maximum demand from the remaining processes and is fed into the grid. Once processes are repaired, the control cycle will be triggered again to increase demand (if own supply remained constant).

Note that above examples aim to illustrate the central control method. Several additional control situations are possible, for example, if a scheduled rate does not equal production rate (blocking or starving, time lags due to mandatory finishing of processing), if continuous processes are rescheduled, which might result in a different (better) fit for binary processes and thus triggering of the central electricity control again or if process self-switch-off is enacted (detail provided in the following). Time intervals between control actions are, in most cases, not equal and can be very frequent, for example, in systems with a high number of processes as each process state change triggers the control algorithm.

In order to extend and evaluate effectiveness of proposed central energy control, two additional control strategies are introduced: energy efficiency control and no control.

The energy efficiency control aims at switching processes off when incoming and outgoing buffer fill levels are adequate to not impact adjacent processes. Time window value $TW_{n,t}$ in relation to switch on and off times ($SOnT_n$ and $SOffT_n$), adjusted by a security factor to account for delayed buffer refill/withdraw (energy saving factor esf_n, to be increased if delayed buffer refill causes throughput impact), is used to determine process switch off. If energy efficiency control is enacted, an idle wait time counter with wait time wt_n is started whenever a process enters idle state (e.g., due to energy control actions or blocking/starving). Idle wait time is similar to compressor wait time before switch off. However, if idle wait time has passed and no control signal to resume production has been received, and the following condition is true,

$$TW_{n,t} \geq esf_n \cdot (SOnT_n + SOffT_n)\,, \tag{4.53}$$

processes are scheduled to be switched off. When available time window becomes lower than required time for switching the process back on (multiplied by the energy saving adjustment factor), the process is scheduled for production and switched on. Further, depending on desired objectives (i.e., energy saving vs. VRE utilization increase), processes can be allowed to be switched on during energy saving off-state if scheduled by the central energy control to increase VRE utilization, or processes can only be allowed to be switched on if buffer levels become critical (to potentially reduce total energy demand).

The second additional control strategy, no control, provides a reference basis to evaluate the impact and effectiveness of other control strategies. Based on a pull-principle (Günther and Tempelmeier 2005), processes start processing of a part when a part has been removed from the process' outgoing buffer. Consequently, when a customer withdraws a part from the outgoing buffer, the last process in the system starts production of a part by withdrawing a part from its incoming buffer, which then triggers processing for the previous process. No energy-related control is enacted.

The next section describes data analysis and evaluation to enable interpretation of model outputs.

4.6 Analysis and Evaluation

The analysis and evaluation module collects model input and output data for storage, further processing, evaluation and visualization purposes (Fig. 4.21). As such, it acts as the interface to the user. Aside from formal data management (D1), three different interface/evaluation modules are suggested: a real-time reporting and visualization level (D2), which allows monitoring model evolution during calculation, a scenario evaluation (D3), which provides aggregated indicator data for a given model evolution and set of inputs (a scenario), and an experiment evaluation (D4), which compares results (indicators) for different scenarios.

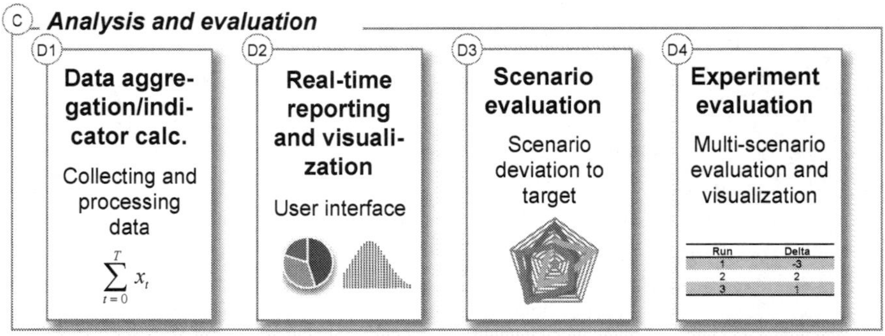

Fig. 4.21 Analysis and evaluation elements (compare also to Fig. 4.4, own illustration)

4.6.1 Data Aggregation and Indicator Calculation

Relevant data from all model elements needs to be collected, processed, and stored within a database to allow detailed analysis and visualization to compare impact from input structure or parameter changes, i.e., different scenarios. Due to the generic nature of this concept, only indicator calculation and general evaluation principles are presented in this chapter. The following chapter, prototypical implementation, will provide detailed information on implementation and visualization options, including data management options. For an overview on production-related data management, i.e., classification into master (permanent) and dynamic data, see, e.g., Schuh and Stich (2012) or Loos (1999).

4.6.2 Real-Time Reporting and Visualization

A central aspect when modeling a complex and dynamic system is to not only generate quantitative results but also to understand the system's behavior (Forrester 1993; Sterman 2000; Meadows 2008). Further, graphical illustration, animation, and visualization in general can be used to validate a model's behavior as the user can collectively observe several system state-describing variables, which can be compared to observed or expected system behavior (Rabe et al. 2008). Real-time reporting and visualization via a GUI is dependent on the chosen model implementation method. Therefore, a specific description and examples will be provided in the following chapters. For general guidelines on GUI design see, e.g., Shneiderman and Plaisant (2005) or Galitz (2007).

4.6.3 Scenario Evaluation

For the purpose of this concept, a scenario describes a future situation with a focus on decisions and how these decisions have (causally) influenced the future state of the system (Mietzner 2009). Multiple scenarios can be compared and deviations investigated to anticipate the effect of actions and to improve a given target achievement (for further detail and an example which utilizes scenario technique in combination with simulation to improve factory layouts see, e.g., Dombrowski and Ernst 2013). For example, investigating the impact of different throughput targets (customer demand rates), multiple scenarios with different customer demand values could be modeled and evaluated. In order to investigate the difference between scenarios, indicators (sometimes referred to as key performance indicators (KPIs)) can be used to describe the difference between scenarios in relation to a target (von Colbe 1969; AWF 2005; Schuh and Stich 2012; Brecht 2012).

A diverse set of financial, production, and other indicators can be defined and calculated (for a detailed summary of production-related indicators, see, e.g., AWF 2005, for a method to develop and examples for energy-related KPIs (e-KPIs) see May et al. 2015 or Javied et al. 2015, who refer to energy performance indicators (ENPIs)). In general, indicators can be defined as needed, but should be meaningful to describe a target achievement and to reveal influencing factors. Based on existing research described in Chap. 3, the concept's target (energy flexibility improvement) and existing classification in literature, four indicator classes are introduced:

- Energy flexibility indicators
- Operational indicators
- Economic indicators
- Ecological indicators.

The next sections provide a brief overview of selected indicators for each class.

Energy Flexibility Indicators

Energy flexibility indicators aim at providing a mean to evaluate how well a system change, control strategy or other potential improvement measure contributes to DR capabilities of the overall system. Most relevant indicators include:

- **Self-sufficiency ratio** SSR_t (in percent), which denotes the share of own electricity demand $CEl^{own}_{sys.,t}$ from total electricity demand $CEl^{total}_{sys.,t}$ until time t, i.e., how much electricity demand was satisfied by on-site generated electricity: $SSR_t = CEl^{own}_{sys.,t}/CEl^{dem.}_{sys.,t}$. This measure is the main indicator to evaluate energy flexibility.
- **Cumulative electricity** CEl_t in watt-hours (calculation according to Eq. 4.3) indicates total electricity demand, supply or feed-in until time t. Different system parts and objects (e.g., multiple processes, compressors) and energy sources (e.g., wind, solar) can be referenced and values can be calculated on an average per product basis by dividing obtained values by total throughput. Further, a product-individual value can be obtained by dynamically allocating electricity on a per

source basis (i.e., if current (at time t) wind share of total electricity demand is 50 % and a given product is processed, required electricity and embodied energy in CA and battery can be attached to a product, compare also to Fig. 4.16).

- **Maximum external demand** MED_t in kW is the highest/peak electricity demand from the system over a given time period. As noted before, either a block average or rolling average can be used, while the maximum rolling average is a more conservative measure as it is at least as high as the maximum block average. Further, without knowing the exact start and end time of block average calculations, using a rolling average has the advantage that a specific start or end time is not required. As such, a rolling average calculation is used in the following and calculated over (typically) 15 min, i.e., $MED_t = \int_{t-15}^{t} El_t^{grid}/15\, dt$ with t measured in minutes, from which a maximum value is determined. A block average calculation is similarly performed using distinct values for time intervals at which the value is obtained, i.e., only for $MED_{15 \cdot x}$ with $x \in \mathbb{N}$. Especially for small companies, peak power demand charges can incur substantial additional costs (see, e.g., Thiede 2012 for further details).
- **On-site generation demand ratio** $OGDR_t$ (in percent), which is defined as the share of cumulative own electricity demand $CEl_{sys.,t}^{own}$ from total on-site generated electricity CEl_t^{own} until time t, i.e., how much of on-site generated electricity was demanded by the system: $OGDR_t = CEl_{sys.,t}^{own}/CEl_t^{own}$. Note that $OGDR_t \in [0, 1]$ and that $1 - OGDR_t$ denotes the share of on-site generated electricity fed back into the grid.

Above description summarizes the central indicators required to determine system performance with regards to energy flexibility. Additional indicators, such as embodied electricity in products, compressed air or on-site generated electricity fed back to the grid should be evaluated and monitored on an as-needed basis. Examples for additional indicators will be provided within the next chapters (implementation and case study).

Operational indicators

Operational indicators evaluate the performance in relation to operational (manufacturing and production) targets. Indicators include direct production-related indicators (e.g., throughput), but also process, quality, and logistic ratios (which are sometimes regarded as separate indicator classes, see, e.g., AWF 2005):

- **Throughput** Tr_t in products/pieces is the sum of final products produced and withdrawn from the last buffer by a (virtual) customer until t.
- **Average inventory** AvI_t is the arithmetic average over time for all (intermediate) product inventories until t.
- **System residence time** Srt_t in, e.g., minutes is the time a product resides in the system (time from entering the system's incoming buffer to exiting the outgoing buffer). Can be calculated as arithmetic mean over all products or on a per product basis to, for example, determine a distribution and confidence intervals.
- **Compressor switches** CSS_t as count is the rolling switch count per hour of an individual compressor. Maximum allowed switches (depending on compressor type)

should not be exceeded to prevent compressor failures and fast wear-out (for an indication on maximum compressor switches see Table 4.2). Can be calculated on an individual compressor basis, maximum values for model time can be evaluated as well as distributions and related statistics (e.g., confidence intervals).

- **Battery statistics** including total cycles $Cyc_{b,t}$ and electrical energy throughput (e.g., $CEl_{b,t}^{out}$) according to Sect. 4.3.4 should be reported to estimate battery lifetime and the battery's role as (intermediate) energy storage option.

Economic Indicators

Economic indicators are used to estimate monetary benefits and costs of a given scenario setup. Proposed concept does not aim at providing a full production/manufacturing company (operating) cost model, but aims at focusing on most relevant cost items. Further, monetary values which remain (nearly) constant, i.e., which are independent of model parameters (e.g., production facility rent) are neglected. Consequently, only differences between scenario evaluations can be used for interpretation (i.e., scenario A is economically preferable to scenario B based on a calculated monetary value difference). This approach has the advantage of requiring less input parameters, which can be hard to estimate/evaluate for a full company cost model.

An overview of how total monetary value MV_t is calculated, including main influencing factors, is provided in Fig. 4.22. Total monetary value is the sum of total inventory holding costs (negative value), external electricity costs (negative value), feed-in compensation (positive value), and embodied electricity value (positive value):

- **Total inventory holding costs** are calculated by determining total inventory value and applying a (quasi) continuous interest rate, e.g., weighted average cost of capital (WACC) for the respective company. Total inventory value at a given point in time is the sum of all individual product values (in buffers and processes). Individual product value is calculated by adding an hourly machine rate to an initial value of an individual product during processing (i.e., an hourly rate of 60 Euro and a processing time of 15 min would add 15 Euros to the product's value). Machine (process) hourly rate needs to be estimated as input parameter for every manufacturing process.
- **External electricity cost** is the sum of costs for electric energy and power demand. Energy costs are calculated by multiplying external (grid) electricity demand with the company's electricity price (e.g., in EUR/kWh). If TOU-rates or RTP is enacted, dynamic cost calculation and integration over time needs to be applied. Costs for power demand are calculated by determining the (current) peak demand as defined per electricity supply contract (e.g., maximum rolling 15-minute average), multiplied by peak demand charges (e.g., in EUR/kW) and multiplied by the share of current model evolution time from peak demand contract billing period (i.e., if model evolution time is 90 days and peak demand charges are evaluated and paid for a 360-day period, a factor of 90/360 = 0.25 is applied). This ensures comparability of different model time horizons.
- **Feed-in compensation** is the benefit for supplying on-site generated electricity to the grid. Total electricity fed to the grid by supply source (wind, solar) is multiplied

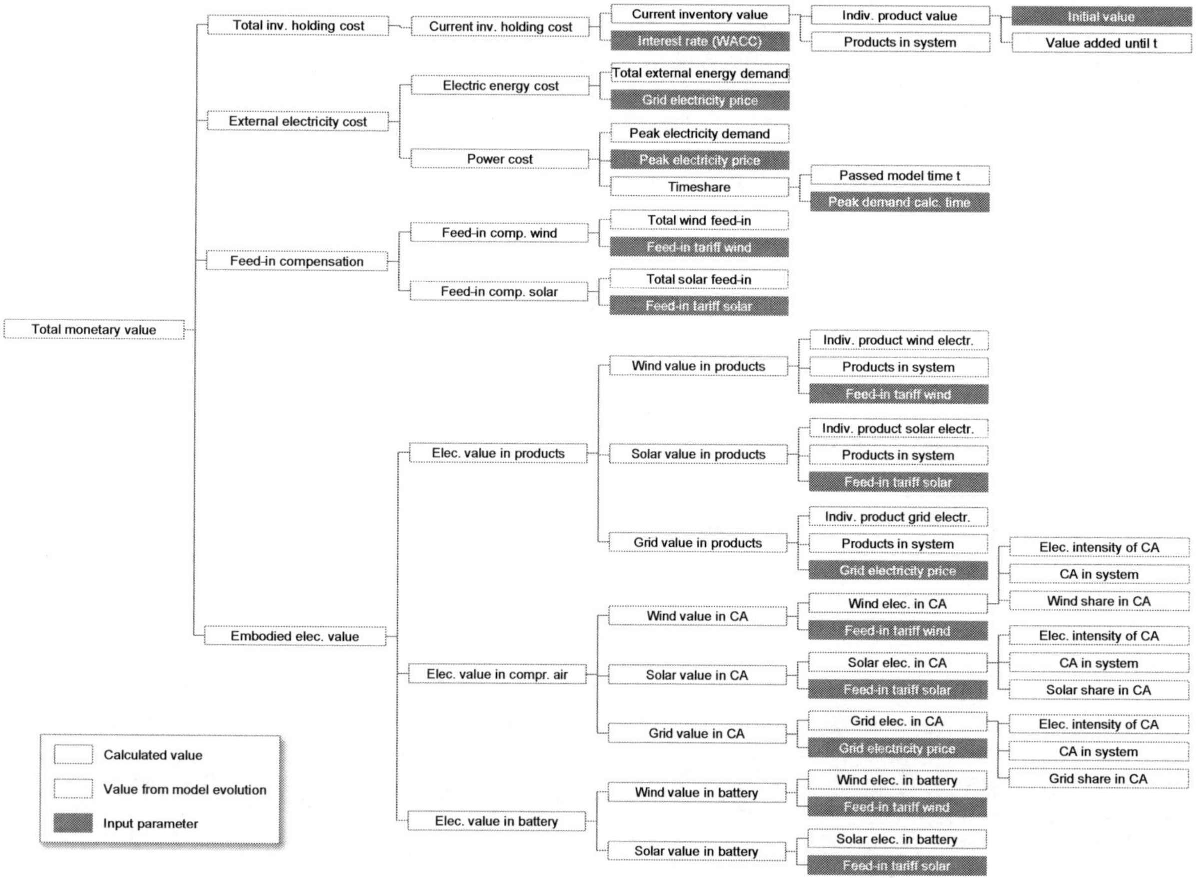

Fig. 4.22 Monetary evaluation driver tree (WACC: Weighted Average Cost of Capital, t: model time, CA: compressed air, own illustration)

by a source-specific feed-in compensation (e.g., in EUR/kWh). Feed-in compensation can be derived from energy exchanges if electricity is sold on a wholesale market or, for example, in Germany, from mandatory feed-in tariffs.

- **Embodied electricity value** is the monetary value for electricity that has been used for product conversion, CA generation or to charge the system's battery and has not yet left the system due to products not leaving the system or due to conversion inefficiencies. This value is added to total monetary value to not penalize a scenario if beginning and ending inventories, CA pressure or battery SOC are different. As such, only the difference between start and current (ending) values are calculated. Embodied electricity differentiated by supply source of all products currently in the system (according to Sect. 4.4.1) is multiplied by feed-in tariffs (wind, solar) and grid electricity price (for embodied grid electricity) and added. For CA, the current amount of embodied electricity by supply source needs to be tracked. This is accomplished by dynamically calculating the electricity intensity of CA (e.g., in kWh/Nm^3) by integrating electricity demand for CA production and deducting embodied electricity in CA which leaves the system (e.g., to perform work in a manufacturing process). This logic is applied for every supply source separately. Multiplying supply share, embodied electricity, and current CA in the system by source with the source-specific tariff (feed-in or grid price) yields the embodied monetary value in CA. For the case of embodied electricity in batteries (SOC), own supply electricity share of SOC per source is dynamically calculated and multiplied by feed-in tariffs.

Above monetary evaluation has the objective to estimate monetary impact of different system setups (scenarios). As mentioned before, only differences and no absolute values are obtained. Within the application cases described in the following, an example can be found which extends this quantitative evaluation with a qualitative discussion involving added costs due to installing a battery.

Ecological Indicators

Ecological indicators are used to evaluate performance in relation to emissions and ecological impact in general. As outlined before, GHG emissions are a central concern for industry and electricity generation. Additional indicators could be implemented to allow a holistic ecological evaluation towards a complete life cycle assessment (LCA, see, e.g., Herrmann 2010 for an overview on steps involved in a LCA). However, due to its relevance, to reduce complexity and to maintain a clear focus of this concept, GHG emissions and carbon dioxide in particular are chosen as indicator.

- **Total CO_2 emissions** in kg are calculated by multiplying total grid electricity demand by a specific CO_2 emission factor. For simplicity, on-site generated electricity is assumed to have no emissions (can be adjusted by applying a source-specific emission factor as needed). As an example for a specific emission factor, average specific CO_2 emissions for domestic grid demand were 0.595 kg CO_2 per kWh electricity in Germany in 2013 (excluding electricity trade balance, i.e., electricity exports and imports were accounted for, value from Icha 2014). For the importance of how specific emission factors are calculated and the significance of using nested average factors instead of (traditional) average emission factors (for the case of the U.S. aluminum industry) see, e.g., Colett et al. (2015). Especially in a context where grid averages are used, but electricity (sub-)grids have limited transmission capacities across (sub-)grid boundaries or electricity grids exhibit significant electricity imports and exports, using an average factor might not be appropriate to provide the necessary detail to allocate emissions. Further, timing differences might need to be accounted for, i.e., demanding grid electricity during day might have another specific (dynamic) emission factor than electricity demand at night if grid supply has, e.g., a large solar share during day. However, data availability might limit the achievable level of detail for emission factors. In addition, applying proposed emission factor logic neglects emissions from other life cycle stages of deployed equipment, e.g., increased product, compressed air or battery storage. Therefore, comparing two different system setups and CO_2 emission values requires careful consideration of potential additional impacts, i.e., increasing VRE utilization might reduce emissions from electricity demand, but, depending on the specific measure that has been applied to achieve this result, additional emissions might be caused (e.g., deploying an additional battery and/or utilizing intermediate product buffer space for embodied energy storage causes different additional emissions). However, to limit complexity of this concept and considering the variety of (individual) different setups, no general logic is included here to account for additional emissions and an (average) emission factor for grid electricity is used.

- **Per product CO_2 emissions** can be calculated by either obtaining an average or product-individual value. An average value is calculated by dividing total emissions by production volume. However, all products are modeled as individual entities, and converted/stored energy in all relevant elements (compressed air, battery, etc.) is tracked by source (see also Fig. 4.16). Consequently, a time-dynamic emission value per individual product can be obtained by tracking emissions that are caused by processing a given product. For example, a product that is processed during high supply of RE, while also compressed air used in production has been generated with mostly RE, gets less emissions assigned than a product which mainly demanded grid electricity for processing. Further, individual values are also beneficial (increased accuracy) if different product types are processed, e.g., with different processing times. The result of individual emission tracking is a list with all products and their respective individual emissions, highlighting which products are "really" green (produced/processed during high RE share considering all time dynamics including intermediate energy storage) and which products rely on traditional (grid) supply. Further, described logic can be applied to all other indicators where useful, for example, assigning individual cost values (which is used in this model to calculate accurate inventory holding costs).

Several indicators are influenced by starting and ending inventory and energy storage levels. For example, total electricity demand over total model time can ceteris paribus (c.p.) be higher if more intermediate products are stored in buffers (more embodied energy in intermediate products) or ending CA pressure level is higher than initial CA pressure. In order to remove the influence of starting and ending product inventory (embodied energy), CA stored in the system and battery SOC, (embodied) energy is tracked on a per source (grid, wind, solar) basis and all relevant indicators are also calculated with removed system ending state influence.

Scenario evaluation can require multiple system evolution calculations if stochastic variables are used to describe component behavior (i.e., stochastic failure and repair of machines). As a result, multiple scenario calculations are required to obtain a probability distribution of (selected) indicators. Depending on system complexity and desired granularity of obtained distributions (e.g., to fit a common probability distribution such as a log-normal distribution and apply significance tests), required time to calculate results can significantly increase. One option to reduce computation time would be to use maximum likelihood estimates of probabilistic input variables. However, due to existing nonlinear behavior of the model, obtained indicator results are potentially not the maximum likelihood estimate for its underlying distribution function (if input variables would have been modeled as stochastic variables).

4.6.4 Experiment Evaluation

In order to compare the impact of different system and/or parameter changes, experiments are carried out to determine system behavior and characteristics (Park 2007). For the purpose of this concept, an experiment is carried out by evaluating and comparing results from different scenario runs. Experiments can either be hypothesis-driven (e.g., "a larger CA storage tank will c.p. increase energy flexibility of the system") or they are used to determine system behavior in general (i.e., "what is the influence of additional intermediate buffer inventory"). Experiment set-up needs to be performed by the user, who aims at investigating the influence of a certain system/parameter change towards enabling and improving energy flexibility (compare to Fig. 4.4, *V Feedback, adjustment towards improvement*). Experiments can be set up according to user's knowledge and experience (hypothesis-driven) or methods for structured target search can be applied (i.e., Design of Experiments (Park 2007), for an overview on experiment design and techniques see, e.g., Rojas et al. 2007).

4.7 Application Cycle

Proposed concept aims at providing an universally applicable method to test and improve energy flexibility of manufacturing systems. As such, it can be applied at various stages of a manufacturing system's life cycle, e.g., during greenfield planning, as part of deploying on-site RE sources, during alteration of existing structures or simply as an energy flexible manufacturing control strategy without (significantly) changing existing structures.

As outlined before, an energy flexibility planner will be the key user of the concept to promote energy flexibility within an organization or subentity (e.g., subsidy or

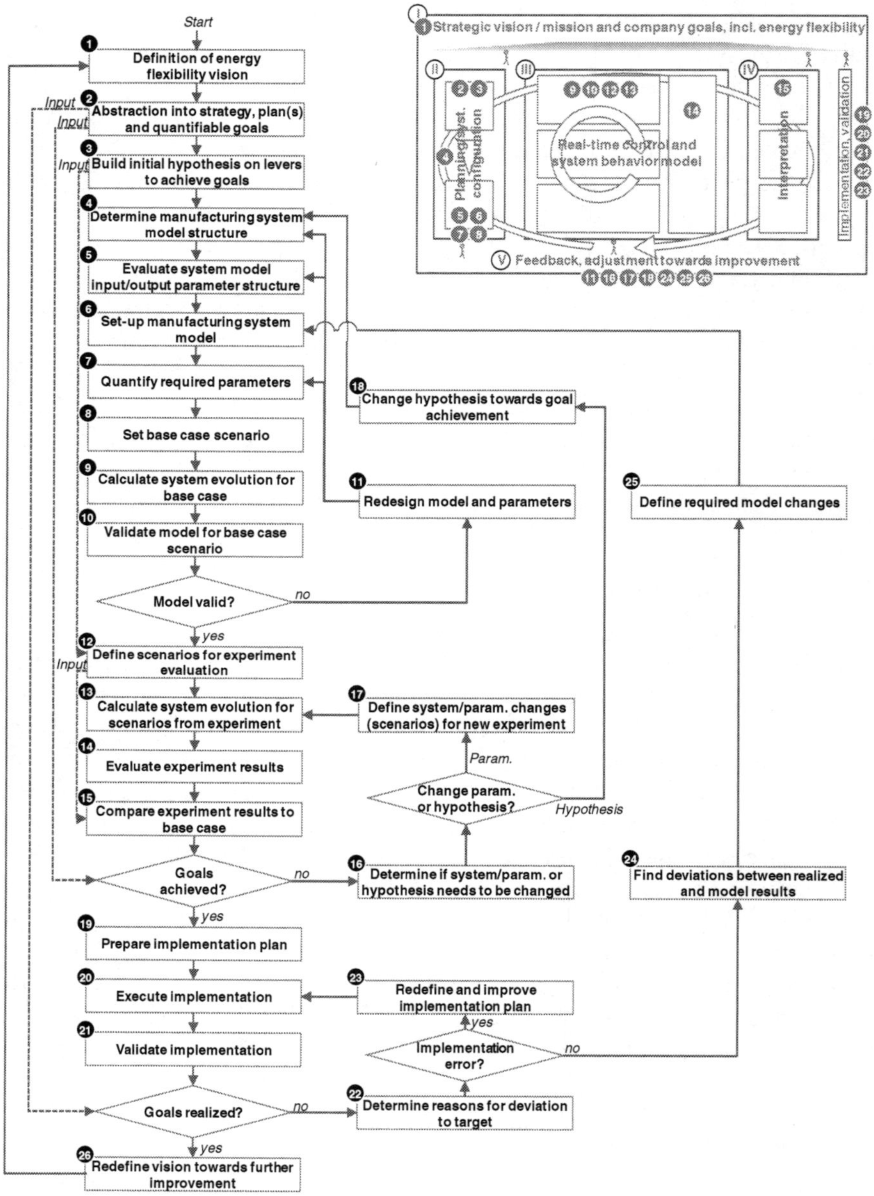

Fig. 4.23 Application cycle steps of proposed energy flexibility improvement concept (own illustration)

factory). Based on the overall concept structure illustrated in Fig. 4.4, an application cycle is described within this section to support application and implementation of proposed concept. The complete application cycle consists of 26 steps and is illustrated in Fig. 4.23. The application cycle includes several feedback loops to ensure validity of the model and results, to improve energy flexibility and to ensure proper implementation. Within the following, tasks and objectives of the different steps are described.

1 Definition of Energy Flexibility Vision

The initial step towards energy flexibility and energy flexible manufacturing systems in particular is integrating an overall energy flexibility vision into the company or organization's vision. This ensures required internal (employees and specifically management) and external stakeholder (e.g., shareholders, customers) attention to enable energy flexibility.

2 Abstraction into Strategy, Plan(s) and Quantifiable Goals

The second step is required to derive measurable indicators to evaluate energy flexibility and its improvement. First, a (short-, medium-, long-)term strategy for energy flexibility needs to be developed according to defined vision. Based on this strategy, corresponding plans to realize the strategy are outlined and indicators are defined that can be used to track success of fulfilling defined plans. These indicators should cover all relevant dimensions, which include energy flexibility and further organizational performance dimensions related to operational, economic, environmental and organizational strategies and targets. Potential indicators are described in Sect. 4.6.3 *Scenario evaluation*. Additional indicators can be defined as needed.

3 Build Initial Hypothesis on Levers to Achieve Goals

Based on targeted goals, an initial hypothesis needs to be formed that defines measures and changes that are expected to yield the desired results. Examples for hypotheses are increased CA system volume for energy storage or increasing intermediate product inventory to improve energy flexibility towards increased demand of on-site generated electricity. The initial hypothesis provides the starting point for structured target search and improvement.

4 Determine Manufacturing System Model Structure

The next step is to abstract the existing real-life system into an appropriate (i.e., according to defined goals and thus required detail), formalized logical structure that can be used to build the quantitative model described in this chapter. The structure should include all relevant information for mathematical system modeling.

5 Evaluate System Model Input/Output Parameter Structure

Based on the formalized description of the manufacturing system structure, all inputs and outputs need to be defined. Input structure is mainly defined by the system model structure. Nonetheless, all defined inputs (parameters) need to be quantifiable, i.e., they must be measurable or obtaining reasonable assumptions based on historic data or expert knowledge needs to be feasible. Output structure is also based on the system's logical structure, but mainly defined by indicator evaluation. All data for indicator evaluation needs to be collected and/or calculated. If either input or output values cannot be obtained or are not quantifiable, system model structure and/or performance indicator structure needs to be adjusted with the goal to find a feasible solution, which still adequately reflects the system's nature and energy flexibility goals.

6 Set Up Manufacturing System Model

After all required input and output parameters and the system's logical structure has been defined, the manufacturing system model is mathematically described. This can either be accomplished by first describing the complete set of equations in textual form, i.e., on paper, or by implementing the model directly into a computerized environment (e.g., available software environments such as Matlab, Plant Simulation, or Anylogic). However, due to the inherent dynamic complexity of the model, a software solution needs to be chosen which allows set up, initial configuration and changes with a reasonable effort while still providing the necessary computational tools and performance to evaluate system evolution. The following Chap. 5 provides more background on model quantitative implementation options and their suitability for proposed concept.

7 Quantify Required Parameters

Defined manufacturing and auxiliary system and further parameters (e.g., electricity generation and price parameters) need to be quantified. For determining manufacturing system parameters, especially energy demand-related values, see, e.g., Posselt (2016) for an overview and detailed description of available methods. Electricity supply data needs to be either estimated from historical values (e.g., time series) or generic functions can be applied (e.g., Heaviside or sine functions) to test the system's response to electricity supply changes. Cost data needs to be determined from, e.g., existing contracts.

8 Set Base Case Scenario

An initial base case needs to be defined to set a baseline for evaluating the impact of changing system parameters and structure for improving desired parameters towards fulfilling energy flexibility objectives. Any base case scenario can be chosen, however, in most cases, the current system setup and state should be selected as it provides the starting point for implementing any improvement measures. For a greenfield planning case or planned system change due to, e.g., product or (production) technology changes, an initially planned system can be used as base case.

9 Calculate System Evolution for Base Case

In order to quantify a baseline for improvement, a system evolution calculation with initial (base case) parameters is performed. Outputs are collected and indicators calculated and stored for comparison to scenario output.

10 Validate Model for Base Case Scenario

An essential step of initial model preparation is to validate the model and its behavior. While verification can also be done during earlier steps (and should be performed as needed), the presented application method suggests to apply verification after initial base case calculation has been performed. This allows to utilize calculated data from calculations to be compared, e.g., to measured and observed data or to allow for consistency and dimensional testing. For an overview on verification and validation methods, see, e.g., Rabe et al. (2008) and Eley (2012). A more detailed discussion can also be found in Sect. 5.2.

11 Redesign Model and Parameters

If validation results indicate that the model is not (sufficiently) valid for its purpose (i.e., calculated results are not in line with expected or observed results), structure and parameters need to be adjusted. Possible changes include redesigning the system formal structure, its input/output parameter set and/or reevaluating values for parameters. Validation and above steps need to be repeated until a valid model is obtained.

12 Define Scenarios for Experiment Evaluation

In the presence of a valid model, scenarios and related input parameters are defined. The previously formulated initial hypothesis can and should guide scenario definition. A complete set of scenarios, i.e., set of different input parameters, collectively forms an experiment for evaluation to test a given hypothesis.

13 Calculate System Evolution for Scenarios from Experiment

Utilizing parameter input from experiment definition, model evolution needs to be calculated for each parameter set. Results are stored in a dedicated database for evaluation and visualization. Depending on model complexity and especially model time horizon, calculation can require significant computational resources and, if time and/or effort are inadequate for desired results, the experiment might need to be redefined (i.e., reducing amount of parameter changes and sensitivity analyses).

14 Evaluate Experiment Results

After model evolution calculations have been performed for all scenarios, indicators are calculated for each scenario. Results can be visualized using, e.g., tables and graphs. Further, additional indicators and ratios can be calculated and visualized to illustrate differences and results from scenario calculations (e.g., percentage change between scenario runs).

15 Compare Experiment Results to Base Case

Obtained experiment results are compared to results from the model base case scenario (and also further evaluated/visualized according to previous step). The initial scenario definition serves as input for interpretation of obtained results. Defined quantitative energy flexibility goals and objectives are compared to results and target fulfillment is determined.

16 Determine if System/Param. or Hypothesis Needs to be Changed

If objectives are not fulfilled, either system/parameter changes (i.e., another experiment setting such as further increasing intermediate buffer capacity) or hypothesis changes (i.e., increasing processing capacity rather than buffer capacity) need to be done. Expert knowledge and findings from previous model calculations are used to guide alteration and target search. Further, structured methods as discussed before (e.g., Design of Experiments) can be applied.

17 Define System/Param. Changes (Scenarios) for New Experiment

If system or parameter changes are determined to improve results towards target fulfillment, a new experiment is set up. The process continues with the new experiment's system evolution calculation.

18 Change Hypothesis Towards Goal Achievement

If experiment changes are considered to not yield desired results or if altering the initial hypothesis as a result of completed experiments is considered more promising than new experiments towards testing the same hypothesis, a new hypothesis is formulated. As this step results in additional modeling and calibration effort (i.e., if the new hypothesis cannot be tested using the current setup), formulating a new hypothesis should be done carefully and only after the initial hypothesis has not been proven a valid instrument to fulfill energy flexibility objectives.

19 Prepare Implementation Plan

If an experiment evaluation indicates that a hypothesis is suitable to achieve energy flexibility goals, an implementation plan is prepared. Necessary changes based on experiment results are outlined. Specific steps, deadlines, and responsibilities need to be assigned. Further, required resources are specified and organization-specific tasks (e.g., project management processes) executed.

20 Execute Implementation

Implementation into the real manufacturing system environment is executed according to the implementation plan. A dedicated person which is responsible for adherence to the initial plan including time frame and budget/resources should be appointed (which can, but does not need to be, the energy flexibility planner/team). An example for implementing energy control and monitoring functions into manufacturing equipment PLCs can be found in Abele et al. (2015).

21 Validate Implementation

Upon completion of implementation, implementation is validated. If feasible, validation takes already place during implementation (i.e., validating results from subsystems and milestones). Actual results are measured and compared to expected results.

22 Determine Reasons for Deviation to Target

If measured and/or observed results from implementation deviate to expected results, potential causes for deviations are investigated (can theoretically be neglected if deviation to target is positive, i.e., a better outcome than expected is realized, which results in losing potential learning/further improvement opportunities).

23 Redefine and Improve Implementation Plan

If the realized deviation from anticipated results is a result of an insufficient implementation or implementation planning, an improved implementation strategy is required. Changes in implementation strategy can, for example, include technical (utilized equipment, software) or organizational (i.e., responsibilities, operating instructions) adaptions.

24 Find Deviations Between Realized and Model Results

If undesired implementation results are a consequence of an insufficient modeling effort, required modeling changes are outlined. The goal is to adjust and redesign the initial model to adequately reflect the system's reality. As the initial modeling effort has not yield desired results and flaws have not been detected during validation and subsequent steps, uttermost care needs to be applied to avoid another unsuccessful modeling process.

25 Define Required Model Changes

Based on identified shortcomings of the initial model, required model changes are determined. Changes are implemented, which results in a step back to setting up the manufacturing system model. As such, redesigning the initial model as a consequence of an unsuccessful implementation potentially results in significant additional effort (time and resources) and should be avoided.

26 Redefine Vision Towards Further Improvement

If implementation is successful and attained results are fully realized, the energy flexibility vision can be revised towards further improvement. This feedback to the beginning of the application cycle ensures a continuous improvement process and thus steady increase in the organizations' energy flexibility performance and related goals (i.e., cost or energy efficiency targets).

4.8 Intermediate Summary and Conclusion

Proposed concept provides an approach to energy flexibility control and improvement of manufacturing systems. Concept objectives have been derived from existing research, which were translated into requirements. Based on these requirements, a concept is proposed that reflects different managerial planning horizons and manufacturing execution in combination. A detailed (mathematical) description of proposed energy flexibility-centered time-dynamic mathematical model and energy flexibility control is given, and complemented by an application cycle for the whole concept.

Although centered around on-site VRE supply, the concept can also be applied for any other limited electricity schedule, i.e., calculated by production planning and control and optimization methods. Proposed concept does not aim at obtaining an optimal dynamic/time-dependent solution. However, OR methods (see Chap. 3 for example methods) can be used to obtain an optimized schedule based on forecasts, while described concept advances OR methods by (longer term) system redesign and real-time energy control to account for forecasting errors and randomness. The next step is a prototypical implementation to test the effectiveness, functionality, and applicability of the concept.

References

Abele, E., Panten, N., & Menz, B. (2015). Data collection for energy monitoring purposes and energy control of production machines. *Procedia CIRP*, *29*, 299–304. http://dx.doi.org/10.1016/j.procir.2015.01.035.

Andrea, D. (2010). *Battery management systems for large lithium ion battery packs*. Norwood: Artech House.

Apple Inc. (2015). Batteries - Why Lithium-ion? Retrieved June 19, 2015, from http://www.apple.com/uk/batteries/why-lithium-ion/.

Aström, K. J., & Murray, R. M. (2008). *Feedback systems: an introduction for scientists and engineers* (2.11b ed.). Princeton: Princeton University Press.

AWF. (2005). *Praktischer Einsatz von Kennzahlen und Kennzahlensystemen in der Produktion*. AWF-Selbstverlag, Eschborn.

Bajpai, P., & Dash, V. (2012). Hybrid renewable energy systems for power generation in stand-alone applications: A review. *Renewable and Sustainable Energy Reviews*, *16*(5), 2926–2939. http://dx.doi.org/10.1016/j.rser.2012.02.009.

Banerjee, S. (2014). *Mathematical modeling: models, analysis and applications*. Boca Raton, FL: Taylor and Francis Group.

Bernal-Agustín, J. L., & Dufo-López, R. (2009). Simulation and optimization of stand-alone hybrid renewable energy systems. *Renewable and Sustainable Energy Reviews*, *13*(8), 2111–2118.

Bertsche, B., & Lechner, G. (2004). *Zuverlässigkeit im Fahrzeug- und Maschinenbau* (3rd ed.). Berlin: Springer.

Bhandari, B., Poudel, S. R., Lee, K.-T., & Ahn, S.-H. (2014). Mathematical modeling of hybrid renewable energy system: A review on small hydro-solar-wind power generation. *International Journal of Precision Engineering and Manufacturing Technology*, *1*(2), 157–173. http://link.springer.com/10.1007/s40684-014-0021-4.

Bierbaum, U., & Hütter, J. (2004). *Druckluft Kompendium* (6th ed.). Darmstadt: Hoppenstedt-Verlag.

Birolini, A. (1999). *Reliability engineering: theory and practice* (3rd ed.). Berlin: Springer.

Brecht, U. (2012). *BWL für Führungskräfte: Was Entscheider im Unternehmen wissen müssen* (2nd ed.). Wiesbaden: Springer Gabler.

Brent, A. C., & Rogers, D. E. (2010). Renewable rural electrification: Sustainability assessment of mini-hybrid off-grid technological systems in the African context. *Renewable Energy*, *35*(1), 257–265. http://dx.doi.org/10.1016/j.renene.2009.03.028.

Chiang, S. Y., Kuo, C. T., & Meerkov, S. M. (1998). Bottlenecks in Markovian production lines: A systems approach. *IEEE Transactions on Robotics and Automation*, *14*(2), 352–359.

Chung, C. A. (2004). *Simulation modeling handbook: a practical approach*. Boca Raton, FL: CRC Press.

Colett, J. S., Kelly, J. C., & Keoleian, G. A. (2015). Using nested average electricity allocation protocols to characterize electrical grids in life cycle assessment: A case study of U.S. primary aluminum production. *Journal of Industrial Ecology*, *20*(1), 29–41.

Deshmukh, M. K., & Deshmukh, S. S. (2008). Modeling of hybrid renewable energy systems. *Renewable and Sustainable Energy Reviews*, *12*(1), 235–249.

Deutsches Institut für Normung e. V. (2011). DIN 22101: Continuous conveyors - belt conveyors for loose bulk materials - basis for calculation and dimensioning.
Dodson, B., & Nolan, D. (2002). *Reliability engineering handbook*. Tucson, AZ: QA Publishing LLC.
Dombrowski, U., & Ernst, S. (2013). Scenario-based simulation approach for layout planning. *Procedia CIRP*, *12*, 354–359. http://dx.doi.org/10.1016/j.procir.2013.09.061.
Duflou, J. R., Sutherland, J. W., Dornfeld, D., Herrmann, C., Jeswiet, J., et al. (2012). Towards energy and resource efficient manufacturing: A processes and systems approach. *CIRP Annals - Manufacturing Technology*, *61*(2), 587–609. http://linkinghub.elsevier.com/retrieve/pii/S0007850612002016.
Ecoult (2015a). Solar Energy Shifting and Power Smoothing. Retrieved October 8, 2015, from http://www.ecoult.com/case-studies/pnm-nm-usa-solar-power-smoothing-and-energy-shifting/.
Ecoult (2015b). Wind Smoothing and Ramp Rate Control. Retrieved October 8, 2015, from http://www.ecoult.com/case-studies/hampton-wind-farm-australia-wind-smoothing/.
Eley, M. (2012). *Simulation in der Logistik: Eine Einführung in die Erstellung ereignisdiskreter Modelle unter Verwendung des Werkzeugs "Plant Simulation"*. Berlin, Heidelberg: Springer.
Forrester, J. W. (1993). System dynamics and the lessons of 35 years. In K. B. De Greene (Ed.), *A system-based approach to policymaking* (pp. 199–240). New York, NY, US: Springer.
Fraunhofer, I. S. I. (2003). Druckluft effizient. Technical report, Karlsruhe.
Galitz, W. O. (2007). *The essential guide to user interface design: An introduction to GUI design principles and techniques* (3rd ed.). Indianapolis, IN: Wiley.
Graßl, M. (2015). *Bewertung der Energieflexibilität in der Produktion*. München: Herbert Utz Verlag.
Graßl, M., Vikdahl, E., & Reinhart, G. (2013). A petri-net based approach for evaluating energy flexibility of production machines. In M. Zäh (Ed.), *5th International Conference on Changeable, Agile, Reconfigurable and Virtual Production (CARV 2013)* (pp. 303–308) Munich, Switzerland: Springer International Publishing.
Guasch, D., & Silvestre, S. (2003). Dynamic battery model for photovoltaic applications. *Progress in Photovoltaics: Research and Applications*, *11*(3), 193–206.
Günther, H.-O., & Tempelmeier, H. (2005). *Produktion und Logistik* (6th ed.). Berlin, Heidelberg, New York: Springer.
Gutowski, T. G., Dahmus, J., & Thiriez, A. (2006). Electrical energy requirements for manufacturing processes. In *13th CIRP International Conference Life Cycle Engineering* (pp. 623–627), Lueven.
Herrmann, C. (2010). *Ganzheitliches Life Cycle Management: Nachhaltigkeit und Lebenszyklusorientierung in Unternehmen*. Heidelberg, Dordrecht, London, New York: Springer.
Herrmann, C., Schmidt, C., Kurle, D., Blume, S., & Thiede, S. (2014). Sustainability in manufacturing and factories of the future. *International Journal of Precision Engineering and Manufacturing Technology*, *1*(4), 283–292. http://link.springer.com/10.1007/s40684-014-0034-z.
Icha, P. (2014). Entwicklung der spezifischen Kohlendioxid-Emissionen des deutschen Strommix in den Jahren 1990 bis 2013. Technical report, Umweltbundesamt, Dessau-Roßlau.
IHK Nürnberg für Mittelfranken. (2012). Druckluft effizient nutzen. Technical report, Industrie- und Handelskammer Nürnberg für Mittelfranken.
Javied, T., Rackow, T., & Franke, J. (2015). Implementing energy management system to increase energy efficiency in manufacturing companies. *Procedia CIRP*, *26*, 156–161. http://linkinghub.elsevier.com/retrieve/pii/S2212827114008701.
Jebaraj, S., & Iniyan, S. (2006). A review of energy models. *Renewable and Sustainable Energy Reviews*, *10*(4), 281–311.
Kara, S., & Li, W. (2011). Unit process energy consumption models for material removal processes. *CIRP Annals - Manufacturing Technology*, *60*(1), 37–40. http://linkinghub.elsevier.com/retrieve/pii/S0007850611000199.
Kuo, C.-T., Lim, J.-T., & Meerkov, S. M. (1996). Bottlenecks in serial production lines: A system-theoretic approach. *Mathematical Problems in Engineering*, *2*, 233–276.

Li, W., & Kara, S. (2011). An empirical model for predicting energy consumption of manufacturing processes: A case of turning process. *Proceedings of the Institution of Mechanical Engineers, Part B: Journal of Engineering Manufacture*, *225*(9), 1636–1646.

Li, L., Chang, Q., Ni, J., Xiao, G., & Biller, S. (2007). Bottleneck detection of manufacturing systems using data driven method. In *IEEE International Symposium Assembly Manufacturing* (pp. 76–81), Ann Arbor, MI. http://ieeexplore.ieee.org/lpdocs/epic03/wrapper.htm?arnumber=4288452.

Li, L., Chang, Q., & Ni, J. (2009). Data driven bottleneck detection of manufacturing systems. *International Journal of Production Research*, *47*(18), 5019–5036.

Li, J., Gee, A. M., Zhang, M., & Yuan, W. (2015). Analysis of battery lifetime extension in a SMES-battery hybrid energy storage system using a novel battery lifetime model. *Energy*, *86*, 175–185. http://linkinghub.elsevier.com/retrieve/pii/S0360544215004843.

Loos, P. (1999). Grunddatenverwaltung und Betriebsdatenerfassung als Basis der Produktionsplanung und -steuerung. In H. Corsten & B. Friedl (Eds.), *Produktionscontrolling* (pp. 227–252). München: Vahlen.

Mason, S. J., Fowler, J. W., & Matthew Carlyle, W. (2002). A modified shifting bottleneck heuristic for minimizing total weighted tardiness in complex job shops. *Journal of Scheduling*, *5*(3), 247–262.

May, G., Barletta, I., Stahl, B., & Taisch, M. (2015). Energy management in production: A novel method to develop key performance indicators for improving energy efficiency. *Applied Energy*, *149*, 46–61. http://linkinghub.elsevier.com/retrieve/pii/S0306261915003578.

Meadows, D. H. (2008). *Thinking in systems: A primer*. London: Sustainability institute.

Mietzner, D. (2009). *Strategische Vorausschau und Szenarienanalysen: Methodenevaluation und neue Ansätze*. Wiesbaden: Gabler.

Mönch, L., & Zimmermann, J. (2007). Simulation-based assessment of machine criticality measures for a shifting bottleneck scheduling approach in complex manufacturing systems. *Computers in Industry*, *58*(7), 644–655.

Müller, E., Engelmann, J., Löffler, T., & Strauch, J. (2009). *Energieeffiziente Fabriken planen und betreiben*. Berlin, Heidelberg: Springer.

Park, G.-J. (2007). *Analytic methods for design practice*. London: Springer.

Paulus, M., & Borggrefe, F. (2011). The potential of demand-side management in energy-intensive industries for electricity markets in Germany. *Applied Energy*, *88*(2), 432–441.

Posselt, G. (2016). *Towards energy transparent factories*. Cham, Switzerland: Springer International Publishing.

Rabe, M., Spieckermann, S., & Wenzel, S. (2008). *Verifikation und Validierung für die Simulation in Produktion und Logistik: Vorgehensmodelle und Techniken*. Berlin, Heidelberg: Springer.

Rojas, C. R., Welsh, J. S., Goodwin, G. C., & Feuer, A. (2007). Robust optimal experiment design for system identification. *Automatica*, *43*(6), 993–1008.

Roser, C., Lorentzen, K., & Deuse, J. (2014). Reliable shop floor bottleneck detection for flow lines through process and inventory observations. *Procedia CIRP*, *19*, 63–68. http://linkinghub.elsevier.com/retrieve/pii/S2212827114006520.

Ruppelt, E. (2003). *Druckluft Handbuch* (4th ed.). Essen: Vulkan-Verlag.

Schuh, G., & Stich, V. (2012). *Produktionsplanung und -steuerung 1: Grundlagen der PPS* (4th ed.). Berlin, Heidelberg: Springer Vieweg.

Schultz, C., Sellmaier, P., & Reinhart, G. (2015). An approach for energy-oriented production control using energy flexibility. *Procedia CIRP*, *29*, 197–202.

Shneiderman, B., & Plaisant, C. (2005). *Designing the user interface: strategies for effective human-computer interaction* (4th ed.). Boston: Pearson/Addison-Wesley.

Shrouf, F., Ordieres-Meré, J., García-Sánchez, A., & Ortega-Mier, M. (2014). Optimizing the production scheduling of a single machine to minimize total energy consumption costs. *Journal of Cleaner Production*, *67*, 197–207. http://linkinghub.elsevier.com/retrieve/pii/S0959652613008780.

Sproedt, A., Plehn, J., Schönsleben, P., & Herrmann, C. (2015). A simulation-based decision support for eco-efficiency improvements in production systems. *Journal of Cleaner Production, 105*, 389–405.

Sterman, J. D. (2000). *Business dynamics: Systems thinking and modeling for a complex world*. Boston, MA: The McGraw-Hill Companies Inc.

Sun, Z., & Li, L. (2014). Potential capability estimation for real time electricity demand response of sustainable manufacturing systems using Markov Decision Process. *Journal of Cleaner Production, 65*, 184–193. http://linkinghub.elsevier.com/retrieve/pii/S0959652613005738.

Thiede, S. (2012). *Energy efficiency in manufacturing systems*. Berlin, Heidelberg: Springer.

United Nations. (2009). *Recommendations on the transport of dangerous goods*. New York, NY Geneva: United Nations Publications.

von Colbe, W. B. (1969). *Kennzahlen und Kennzahlensysteme als Mittel der Organisation und Führung von Unternehmen*. Wiesbaden: Springer.

Wiendahl, H.-P. (2014). *Betriebsorganisation für Ingenieure* (8th ed.). München: Hanser Verlag.

Wilker, H. (2010). *Weibull-Statistik in der Praxis: Leitfaden zur Zuverlässigkeitsermittlung technischer Komponenten* (2nd ed.). Norderstedt: Books on Demand GmbH.

Yang, H., Lu, L., & Zhou, W. (2007). A novel optimization sizing model for hybrid solar-wind power generation system. *Solar Energy, 81*(1), 76–84.

Yang, H., Wei, Z., & Chengzhi, L. (2009). Optimal design and techno-economic analysis of a hybrid solar-wind power generation system. *Applied Energy, 86*(2), 163–169.

Yusta, J., Torres, F., & Khodr, H. (2010). Optimal methodology for a machining process scheduling in spot electricity markets. *Energy Conversion and Management, 51*(12), 2647–2654. http://linkinghub.elsevier.com/retrieve/pii/S0196890410002098.

Zhai, Y., Sun, S., Wang, J., & Niu, G. (2011). Job shop bottleneck detection based on orthogonal experiment. *Computers and Industrial Engineering, 61*(3), 872–880.

Zhang, Y., Hu, W., Chen, Z., & Cheng, M. (2013). Mitigation of wind power fluctuation by active current control of variable speed wind turbines. *International Journal of Smart Grid and Clean Energy, 2*(2), 252–257.

Zhou, Z., & Li, L. (2013). Real time electricity demand response for sustainable manufacturing systems considering throughput bottleneck detection. In *IEEE International Conference on Automation Science and Engineering* (pp. 640–644). Madison, WI.

Zhou, W., Lou, C., Li, Z., Lu, L., & Yang, H. (2010). Current status of research on optimum sizing of stand-alone hybrid solar-wind power generation systems. *Applied Energy, 87*(2), 380–389. http://dx.doi.org/10.1016/j.apenergy.2009.08.012.

Chapter 5
Prototypical Implementation

The concept and mathematical models developed in the previous chapter have been implemented into a working software prototype to evaluate its applicability, usability, and contribution towards integration of variable electricity supply into manufacturing systems. Further, the prototype resembles a universally applicable, computerized calculation environment to support application according to proposed application cycle. The chapter starts with a brief discussion on implementation options and their potential applicability to develop a prototype. Simulation is chosen as a suitable method for implementation, and consequently additional background on simulation methods and validation is provided. The next section describes the developed software prototype, including its functionality, modules, inputs, and outputs as well as validation examples. The chapter closes with an intermediate conclusion.

5.1 Implementation Options and Background

As outlined in the concept's requirements and objectives, a focus is set towards easy configuration, visualization, and application of proposed concept. While the proposed mathematical model outlines the logical structure for the real-time control and system behavior model, and Sect. 4.7, including Fig. 4.23, describes an application cycle for proposed concept, a software prototype is developed to provide a tool for application of the concept and the mathematical model in particular. With reference to the application cycle, steps involving calculation of system evolution over time (step nine and thirteen) can be facilitated if an easily configurable, ready-to-use calculation option exists.

In general, to evaluate a given hypothesis and therefore experiments and their scenarios, model evolution under a given set of input parameters needs to be calculated. As such, different calculation methods exists. Based on review of existing

J. Beier, *Simulation Approach Towards Energy Flexible Manufacturing Systems*, Sustainable Production, Life Cycle Engineering and Management, DOI 10.1007/978-3-319-46639-2_5

research and principles of mathematical modeling in general, the following selected implementation options could be considered:

- **Analytical solution**: Given that proposed model is a set of equations, an analytical solution could be pursued. The set of equations from Chap. 4 could be manipulated to obtain an expression for each relevant indicator (i.e., production volume, self-sufficiency). However, due to nonlinearities of proposed model, combined with high-order differential equations (dependent on specific system structure), an analytical solution is unlikely to be obtainable (Sterman 2000).
- **Markov Decision Process (MDP)**: Proposed mathematical model structure could be transformed to an MDP model structure. For example, power control actions of machines could be chosen as the set of (optimal) actions that need to be obtained by solving the MDP problem. Existing mathematical description needs to be modified into an expression for states, state transitions, actions, and rewards. Explicit consideration of probabilistic state transition in MDP modeling makes the approach particularly interesting for stochastic modeling of input parameters, e.g., machine breakdowns or variable (random) processing times. Dynamic programming might be used to solve the MDP problem. For examples, including approaches discussed in Chap. 3 (research review) and references to further MDP modeling approaches see, e.g., Sun and Li (2014), Li and Sun (2013), Du et al. (2015). Nonetheless, transforming proposed mathematical model into an MDP model would require substantial additional effort and reflection of its logical structure is unlikely to be feasible.
- **Artificial Neural Networks (ANN)**: ANN are used to predict behavior of a system based on a given set of input parameters and trained behavior without necessarily knowing its underlying logical structure. As such, an ANN needs to be trained by using input data (sets) and related (realized/observed) output data (sets). As the described mathematical model in Chap. 4 explicitly formulates a logical structure to evaluate behavior and outcomes without the necessity of (observed) output data, ANN are rather unsuitable to solve presented approach. However, ANN could be used if another implementation option has been applied to generate output data. If, for example, the other chosen implementation option requires extensive resources (e.g., computation resources/time), an ANN structure could be developed, trained with generated data and used to evaluate new/more scenarios with the goal to utilize less resources. However, ANN design and computation itself is rather complex and requires extensive resources and therefore proposed procedure needs to be carefully evaluated before implementation. For an example to apply ANN for forecasting grid loads see, e.g., Fiedler et al. (2007) and for short term electricity price forecasting see, e.g., Yamin et al. (2004).
- **Simulation**: Simulation is a widely used technique to determine system or process evolution over time (Kuhn et al. 1993; Banks 1998; Sterman 2000; Chung 2004). Especially in a dynamic context where analytic solution techniques reach a limit, simulation is used to determine system behavior and evolution. Further, simulation is used to generate deeper understanding of a system and its characteristics, including testing of hypothesis to improve performance of the system. In addition,

in most cases, simulation has the advantage of not directly impacting a real-life system and to be more cost efficient as, for example, testing using real-life models or the actual system. Examples for simulation approaches used in manufacturing engineering and energy-related modeling in particular see, e.g., Schönemann et al. (2015), Thiede (2012), Herrmann et al. (2011) and Hesselbach (2012). Reviewed approaches in Chap. 3 and their related approaches applying simulation include Li et al. (2007, 2012a, b). For a comprehensive overview on discrete-event simulation publications including 317 references see Negahban and Smith (2014) and for a review on utilizing discrete-event simulation and system dynamics for logistics and supply chain management see, e.g., Tako and Robinson (2012). For a summary of state of the art on manufacturing system modeling and simulation with a focus on environmental aspects and future research perspectives see, e.g., Thiede et al. (2013).

Based on above discussion of implementation options, simulation is chosen for implementation. Consistent with the definition of a model given on page 116, a model is created to fulfill a certain purpose. Among others, a central goal of a modeling process can be to perform simulations instead of real-life experiments, e.g., to save costs and time (Banerjee 2014). As such, and due to the explicit mathematical formulation and its dynamic nature of proposed mathematical model, simulation to test hypothesis and thus scenarios is applied for prototypical implementation.

5.2 Background on Simulation

Simulation can be defined as "the imitation of the operation of a real-world process or system over time" (Banks 1998, p. 3). As such, several different simulation techniques exist for different application areas. Selected techniques and verification and validation of simulation models are discussed briefly within the following.

5.2.1 Simulation Modeling

In general, the purpose of a simulation might be described as conducting experiments and obtaining (measured) results from a system model which adequately reflects characteristics and behavior of the real system (Mattern and Mehl 1989). More specific, simulation modeling can follow one or more different objectives/purposes, which might include (Kuhn et al. 1993; Chung 2004; Sterman 2000):

- Studying behavior of a system and developing knowledge about its operation.
- To develop new methods, tools, and approaches which are not yet existing (in real life applications).
- Analytical methods are not applicable or not applicable anymore (i.e., due to complexity challenges).

- Performing tests, experiments and measuring results without actual interference with the real system.
- Testing system changes (including, e.g., operating procedures) before applying changes at the real system.
- Increase efficiency and effectiveness (i.e., performance) of the system.
- Mental and analytical capability of humans are not sufficient to understand or analyze a system.
- Conducting experiments for systems which are not possible (e.g., due to ethical issues) or too costly in real life.
- Test possible solutions to (assumptions of) a real-world problem within reasonable time as opposed to testing at a real system with feedback times that are too slow to test different assumptions within adequate time.

Simulation modeling in general is a broad field and encompasses a variety of approaches, methods and tools. For example, process chains can be simulated using a specific Finite Element method to obtain heat distribution patterns in workpieces between different manufacturing steps (Afazov 2013) or process chains can be simulated using state-based behavior of manufacturing processes and thermodynamic models of (connected) auxiliary systems to obtain a factory-wide energy demand pattern and to evaluate energy efficiency improvement potential (Thiede 2012). Advantages of simulation include expanding and compressing time, advanced observation possibilities, and thus diagnosis and exploring possibilities, while disadvantages include that special knowledge for creating (appropriate) simulation models is required and output data and its relation to system characteristics can be hard to interpret (for an overview of advantages, disadvantages and application of simulation see, e.g., Banks (1998)).

Different simulation methods exist, among them discrete and continuous methods (compare to Fig. 5.1). Continuous methods assume that system state changes occur continuously over time, which is, for example, the case for (continuous) energy flows (Mattern and Mehl 1989). As analytical solutions to continuous time formulations are rarely obtainable (e.g., solving differential equations), numerical

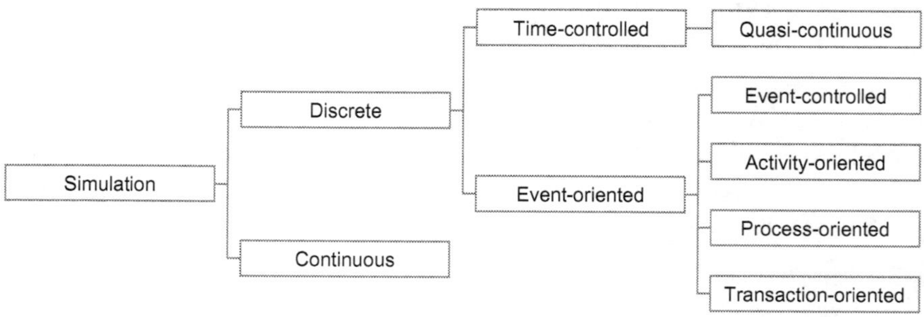

Fig. 5.1 Classification of simulation methods (re-use with permission of Springer from Mattern and Mehl 1989, p. 200, own illustration and translated from German)

methods are frequently applied, causing a simulation to become quasi-continuous. In contrast, discrete-event-based simulations are controlled by events occurring at a defined point in (simulation) time. Model state changes immediately between two events, independent how far simulation time is advanced (in contrast to continuous simulation with a continuous, constant advancement of simulation time). Examples for discrete-event simulation would be processing of a discrete part in a manufacturing machine, with one event describing the *beginning-state* and one event describing the *end-state* of processing the part. For further detail on simulation methods see, e.g., Mattern and Mehl (1989) or Chung (2004).

With the ability to easily combine different simulation methods through available software tools and increasing detail, especially in manufacturing system modeling, multi-method simulation environments are frequently used for a variety of objectives in the context of manufacturing system material and energy flow simulation (for examples, see, e.g., Thiede 2012; Schönemann et al. 2015 and Liraviasl et al. 2015).

5.2.2 Verification and Validation of Simulation Models

Verification and validation are essential steps to ensure that a model behaves in its desired way and adequately reflects the modeled (real) system. The mathematical model described in previous chapters is implemented into a software simulation prototype and therefore different verification and validation methods are required and applied. Consequently, necessary theoretical background is provided here.

On the one hand, verification aims at obtaining a formal proof that the simulation model behaves according to its specifications, i.e., "[...] building the model right" (Balci 1998a, p. 41). Validation, on the other hand, tries to establish a sufficient proof that the model adequately reflects the modeled reality and resembles the original system well enough for its purpose, i.e., "[...] building the right model" (Balci 1998a, p. 41). For a summary of different definitions for verification and validation see, e.g., Rabe et al. (2008). In the context of above description, verification of a model can be done by, e.g., testing its behavior with a set of different input parameters and comparing obtained results with specified (calculated) results. For an overview of verification and validation techniques see Table 5.1, for their detailed description Rabe et al. (2008).

The process of verification and validation is, for most cases, to a certain degree subjective (Balci 1998b; Sterman 2000; Rabe et al. 2008). Some authors even argue that verification and validation of models is impossible (Sterman 2000). In fact, some formal techniques such as formal correctness proofs are only applicable to a very limited number of models which are not too complex or to certain parts of models (Rabe et al. 2008; Balci 1998b). As such, verification and validation are always to a certain degree subjective, as applied tests and techniques need be (subjectively) deemed sufficient to label a model *verified* or *validated*. Consequently, a number

Table 5.1 Overview of techniques used for verification and validation, especially in the context of simulation modeling (according to Rabe et al. 2008)

Nr.	Name	Nr.	Name
1.	Animation	11.	Internal validity test
2.	Cause-effect graph	12.	Monitoring, operational graphics
3.	Comparison to other models	13.	Predictive validation
4.	Desk checking	14.	Review
5.	Dimensional consistency test	15.	Sensitivity analysis
6.	Event validity test	16.	Statistical techniques
7.	Extreme-condition test	17.	Structured walkthrough
8.	Face validity	18.	Submodel testing
9.	Fixed value test	19.	Trace analysis
10.	Historical data validation	20.	Turing test

of techniques outlined in Table 5.1 are applied during development of the modeling process, but without attempt of a formal verification or validity proof.

5.2.3 *Software Packages*

For an overview on available commercial simulation packages and their evaluation in the context of (environmentally focused) manufacturing system modeling see, e.g., Thiede et al. (2013). Commonly used software packages for manufacturing system simulation include, for example, Plant Simulation (Siemens 2015) from Siemens, the discrete-event simulation tool ProModel (ProModel 2015) and the multi-method simulation suite provided by Anylogic® (The Anylogic Company 2015). For this prototypical implementation, the java-based modeling environment Anylogic® has been chosen.

5.3 Mixed Discrete-Event and Continuous Time Simulation Model

The developed prototype was implemented in a mixed discrete-event and continuous time simulation model in Anylogic®. This section therefore outlines selected model elements, including parameters, interfaces to other elements, functionality and visualization. Validation of model elements and the model's overall functionality, including interaction of elements/agents has been done by applying techniques from Table 5.1, e.g., animation, dimensional consistency test, fixed value test, extreme-condition test, monitoring, desk checking, sensitivity analysis, structured walkthrough, and trace analysis. Further, so-called *check-values* are introduced, which are dynamically, i.e., during the simulation run, calculated, and reported as part of an output data(-set). For example, total system cumulative energy demand needs to equal total energy inflows (own supply and onsite generation) minus stored energy (CA, battery) at all times. The total absolute difference can be monitored and any deviation recorded (should be zero, but numerical methods can cause slight deviations). Defining a set of control values contributes to constant verification of the model under a given set of input parameters.

The described mathematical model in Chap. 4 exhibits both continuous and discrete variables and calculation, state-based behavior of modules and several submodels. Anylogic® supports several different modeling techniques through dedicated libraries (e.g., state-based behavior, agent-based modeling), which are used to implement the mathematical logic. In particular, submodels (e.g., batteries, compressors, manufacturing processes, buffers, products) are modeled as agents, which interact with each other. Within the following, the manufacturing process agent type implementation is presented in detail, followed by detail on the energy control and implementation of the user interface. However, to limit the size of this document, not all model details are presented here (e.g., buffer agents, battery agent, data collection module).

5.3.1 Manufacturing Process

The mathematical and logical structure of the manufacturing process is given in Sect. 4.4.1. First, the implementation into an agent type is discussed, followed by its verification.

Implementation

Figure 5.2 shows the implementation of the manufacturing process agent into the simulation environment (for illustrative purposes only, a description of the different elements is provided in the following). Based on its mathematical and logical definition, the agent consists of the following main elements:

- **Material flow logic**: The process collects input products from a connected buffer for processing and sends a processed product to a connected downstream buffer. Product conversion affects energy demand of the process and is dependent upon energy control actions. Further, product characteristics such as current monetary value and embodied electricity by source need to be adjusted according to the process' characteristics (e.g., machine hour rate, current energy demand). In order to design a flexible modeling environment, the process agent has an input and output port which ensures connectivity to other manufacturing system elements through transferring product-agents, as well as a *product sink* for bad parts (if the process fails during product conversion). In Fig. 5.2, the upper left quadrant

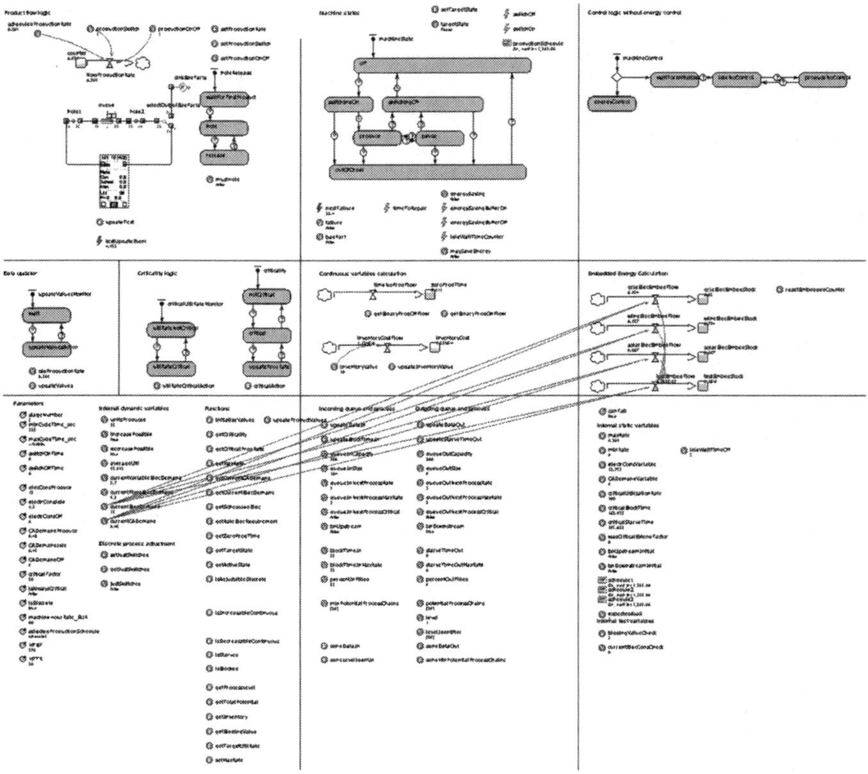

Fig. 5.2 Snapshot of the production process agent's prototypical implementation into a software simulation model (for illustrative purposes only, a description of included elements is provided in this section)

mainly resembles the material flow logic.

- **State-based behavior**: Manufacturing process electricity and CA demand are state-dependent. Consequently, a state chart is used to control energy demand of a process (top row, second element from left in Fig. 5.2). Process states are assumed according to energy control actions (e.g., scheduled process state), material flow logic (e.g., production cannot be interrupted during processing), and failure/out-of order. Each state results in a corresponding energy demand, and only selected available transitions to other states (e.g., from state *out-of-order*, only the state *off* can be reached). Further, transition times such as switch-on/-off are implemented.
- **Energy flexibility control**: Energy flexibility control is realized by implementing a set of functions which determine controllability of the process and which provide an interface to the central control functions to (re-)schedule the process. For example, functions exist which evaluate if the process can be controlled (i.e., critical/noncritical, out-of-order) and what the current electricity demand adjustment potential is. This information is then fed back to the central control logic. Further,

the process monitors if it is throughput critical and what required actions (switch to production) are. In addition, discussed one-piece flow logic is implemented (third element from the left, top row in Fig. 5.2).

- **Data connectivity**: Data connectivity to enable process communication is also realized by providing a data port for connection to other processes (via buffer data ports). A data package containing, e.g., the process' maximum production rate or remaining production capacity is send to upstream and downstream processes. In turn, the process receives data from incoming and outgoing buffers and processes, e.g., to determine if the process is throughput critical and/or impacting a throughput critical process. Further, data connectivity dynamically determines if the process is part of a parallel process chain and subsequently combined throughput potential to determine the bottleneck of a parallel flow system.

Table 5.2 summarizes input parameters to configure the manufacturing process agent. In addition to listed 23 input parameters, dynamic variables, especially data packages for communication between process agents and to enable energy demand and control, are implemented. Further, several internal functions and visualization are part of the agent (compare also to Fig. 5.2).

Verification

Verification of the manufacturing process model is performed by implementing a single process under no energy flexibility control and predefined product demand. Example process parameters can be found in Table 5.2 under P1a. Simulation time is set to 30 min. Products from the process' outgoing buffer are withdrawn with a cycle time of 120 s (equivalent to a rate of one product every 2 min) until 20 min have been passed, after which no additional products are withdrawn. The process is allowed to switch itself off after an idle waiting time of 5 min. The process replaces a product in its outgoing buffer if the buffer level falls below its initial level.

As a central component, energy demand of the process is chosen for example verification. Figure 5.3 shows electricity and CA demand for the simulation time. The process is initially switched on, which requires 3 min. Outgoing buffer level is already 2 pieces below its initial level (withdrawal at $t = 0$ and $t = 2$), the process consequently starts production. Production lasts until $t = 7$ to replace a total of 4 products. Starting at $t = 7$, the process switches from idle to production and back to idle on a per minute basis to replace one product every 2 min until $t = 21$, after witch no products are withdrawn from the buffer. After 5 min idling, the process switches itself off at $t = 26$, and is off at $t = 29$.

Total electricity demand is (compare to Sect. 4.4.1 and Eq. 4.3):

$$\begin{aligned} CEl^{dem.}_{P1,30} &= \int_0^{30} El^{dem.}_{P1,t}\, dt \\ &= 3\text{ min} \cdot 4\text{ kW} + 4\text{ min} \cdot 6\text{ kW} + 7 \cdot (1\text{ min} \cdot 2\text{ kW}) \\ &+ 7 \cdot (1\text{ min} \cdot 6\text{ kW}) + 5\text{ min} \cdot 2\text{ kW} + 3\text{ min} \cdot 1\text{ kW} \\ &+ 1\text{ min} \cdot 0\text{kW} = 105\text{ kW} \cdot \text{min} \end{aligned} \tag{5.1}$$

Table 5.2 Manufacturing process input parameters and four sets (P1a, P1b, P2, P3) of example process parameters (dmnl.: dimensionless)

Nr.	Unit	Description	P1a	P1b	P2	P3
1	Nm^3/min	CA demand during idling	0.2	0	0	0
2	Nm^3/min	CA demand during off-state	0.0	0	0	0
3	Nm^3/min	CA demand during production	0.3	0	0	0
4	Nm^3/min	CA demand during switching off	0.05	0	0	0
5	Nm^3/min	CA demand during switching on	0.05	0	0	0
6	boolean	Switch to determine if process can randomly fail	false	false	false	false
7	dmnl.	Factor to adjust critical block and starve times (until process becomes throughput critical)	10	10	10	10
8	kW	Electricity demand during production	6	6	6	6
9	kW	Electricity demand during idling	2	2	1	3
10	kW	Electricity demand during off-state	0	0	0	0
11	kW	Electricity demand during switching off	1	1	1	1
12	kW	Electricity demand during switching on	4	4	4	4
13	boolean	Switch to set process always into throughput critical state	false	false	false	false
14	boolean	Switch to set process to continuous (false) or discrete (true) process type	true	true	false	true
15	EUR / hour	Machine rate used to determine product value added	60	60	60	60
16	s	Maximum cycle time required for processing	$+\infty$	$+\infty$	$+\infty$	$+\infty$
17	s	Minimum cycle time required for processing	60	60	120	120
18	min	Mean time between failure of the process	N/A	N/A	N/A	N/A
19	min	Mean time to repair the process	N/A	N/A	N/A	N/A
20	dmnl.	Indicates during which centrally provided shift schedule the process can be switched on	S1	S1	S1	S1
21	dmnl.	Processing stage to identify process to a product	1	1	2	3
22	min	Time to switch the process from off to idle/waiting	3	3	3	3
23	min	Time to switch the process from idle/waiting to off	3	3	3	3

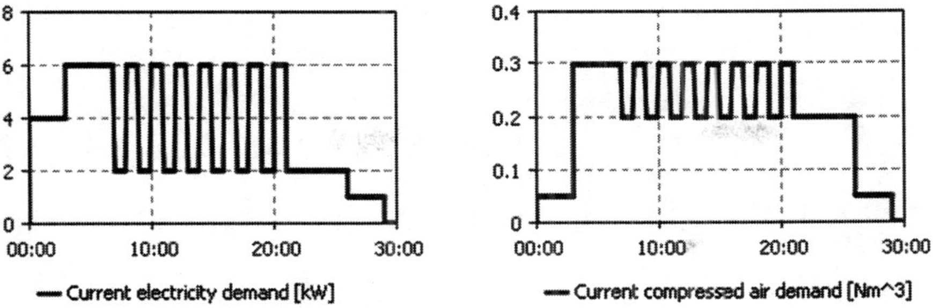

Fig. 5.3 Simulation model output graphs, for example, process electricity demand and CA demand

Correspondingly, total CA demand over 30 min is 6 Nm^3. Cumulative electricity demand calculated by implemented simulation model is 105.058473 kW· min (6.004215 Nm^3), which corresponds to a relative numerical accuracy of 99.944 % (99.930 %). Further, the example illustrates that the process' state-based behavior and its self-switch-off functionality works correctly. Total product output is eleven pieces, as expected.

5.3.2 *Energy Control*

Central energy control is implemented as described in Sect. 4.5, using a set of state-charts and functions. State changes are triggered if an energy control action is required. Energy control is realized by a set of programmed functions (source code rather than, e.g., state-charts or preimplemented program library elements) which determine the desired production control actions to increase utilization of onsite generated electricity.

Functionality-testing and example verification of the central electricity control is accomplished by modeling a simple three-step process chain with intermediate buffers and evaluate the energy control's reaction to changing reference values, i.e., different onsite electricity generation values. Parameters for the three sequential processes are provided in Table 5.2 and labeled P1b, P2, and P3. All processes have no CA requirement. Further, electricity demand values are slightly different. The second process (P2) is modeled as continuous process. With regards to cycle times, the first process requires one minute for processing, i.e., has a rate of one product per minute, while the second and third process' maximum rate is 0.5 products per minute. Total simulation time is 45 min, own electricity supply values change every 10 min.

Figure 5.4 summarizes input parameters and results for the given example. The graphs illustrate the following values over the simulation time:

- Total (system) own direct and grid electricity demand (a), i.e., how much onsite available electricity is directly demanded and how much system demand was satisfied by grid supply.

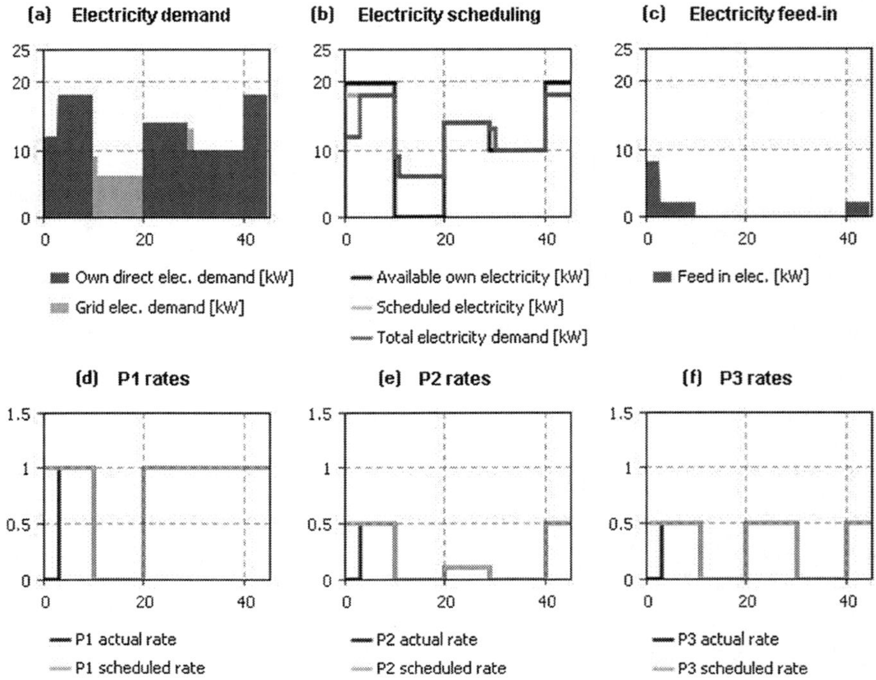

Fig. 5.4 Simulation model output graphs for electricity control example

- Total own (onsite) generated electricity, scheduled electricity and total actual electricity demand (b). Ideally, all lines are matching each other: scheduled electricity demand indicates what demand the energy control aims to achieve, while total electricity demand is the resulting, realized/actual electricity demand. Scheduled electricity demand can deviate from own electricity supply due to limited control options, while total electricity demand can deviate from scheduled electricity if processes cannot achieve scheduled demand due to missing parts (starvation), blocking or if the process needs to be switched on before scheduled (processing) state can be assumed.
- Resulting electricity feed-in (c), i.e., positive net difference between own supply and total electricity demand.
- Scheduled and actual production rate of process one to three (d–f), to retrace electricity control actions.

In total, own electricity demand changes four times, which causes different central electricity control actions. Combined with system material flow and energy demand dynamics, the following behavior can be observed:

- $0 \leq t < 3$: Onsite generation is 20 kW, consequently all processes are scheduled for production (total scheduled electricity demand is 18 kW). However, processes are switched off at $t = 0$, and require until $t = 3$ to be switched to production, resulting in an actual demand of 12 KW and 8 kW feed-in.
- $3 \leq t < 10$: Processes are switched on and assume their processing state (with maximum rate for continuous process P2), resulting in 18 kW of electricity demand and 2 kW feed-in.
- $10 \leq t < 11$: Own supply is reduced to zero at $t = 10$. Consequently, all process's rates are scheduled to zero. However, only processes P1a and P2 can assume idling immediately (2 kW and 1 kW idle electricity demand, respectively). Process P3 is currently converting a product (started at $t = 3$ with a cycle time of 2 min, therefore started conversion of the forth product at $t = 9$) and can therefore not switch to idling, adding another 6 kW of electricity demand.
- $11 \leq t < 20$: P3 has finished processing its part, consequently all processes are switched to idling, requiring a total of 6 kW supplied by external grid sources.
- $20 \leq t < 29$: Own supply is increased to 14 kW. Consequently, both discrete processes are scheduled for production (total demand of 12 kW) and P2, the only continuous process, is scheduled to a rate of 0.1 products per minute, resulting in 2 kW electricity demand. Consequently, the system demands a total of 14 kW electricity, matching own supply.
- $29 \leq t < 30$: Own supply is reduced to 10 kW. The electricity control determines that leaving P1 in production mode and switching P3 to idle yields the closest fit (9 kW) to own supply. The remaining required fit is performed by rescheduling the continuous process, in this case to a zero production rate. However, P3 needs to finish processing of the current part, resulting in a delayed adjustment.
- $30 \leq t < 40$: P3 has assumed idling, resulting in a total system demand of 10 kW and thus matching own supply.
- $40 \leq t \leq 45$: Own supply is increased to 20 kW. All processes are scheduled for production, resulting in 18 kW system electricity demand and 2 kW feed-in.

As the example and description illustrates, the electricity control reacts as expected to electricity supply changes. Additional testing, especially utilizing more complex process chains, including testing for correct bottleneck detection, and different own supply functions (i.e., Heaviside and sine functions) has been performed to ensure correct functioning of the simulation prototype.

5.3.3 Additional Agents and Modules

Additional agents include, among others, a product agent, the compressor park with its pressure control and multiple compressors as well as energy-by-supply-source tracking and a battery park with multiple batteries and energy supply tracking. All functions, interfaces and especially dynamic behavior have been tested for their correct behavior and functionality.

A special focus has been set on easy, fast, and safe set-up, configuration and adjustment of a model. Therefore, elements can be added and removed freely to a model, i.e., a production process can be added to the system by simple drag-and-drop and connecting of ports, a compressor is simply added to the compressor park by drag-and-drop. The newly created agents are then parameterized as desired. All functions, including material and energy flow logic, as well as control functions (e.g., pressure control, energy flexibility control) are automatically adjusted and reflect the presence (or absence in case of deleting an agent) of the new agent in their behavior.

5.3.4 User Interface

Solutions for different application cases can be implemented in Anylogic® by modifying the general, comprehensive user interface structure briefly outlined in this section.

Input Data

Anylogic® offers the opportunity to export and import data from data bases, including files generated by Microsoft Excel and Access, structured text files, and Microsoft SQL Server. Due to limitations inflicted by different licenses (e.g., limited connectivity options to MS Access databases under academic licensing and thus reduced potential user group of the prototype) and widespread use, Microsoft Excel was chosen for input configuration. This includes, for example, solar and wind generation data and product characteristics such as cycle times and corresponding energy demand by production stage. Other, especially static and one-dimensional parameters (e.g., grid electricity price) are directly manipulated with AnyLogic®. Scenarios and experiments can easily be set-up using Anylogic®'s parameter variation function. An advantage of utilizing this feature is Anylogic®'s support of multi-threading, i.e., parallel execution of model instances with different input parameter sets using multi-thread and multi-core capabilities of current computer hardware systems, which significantly increases computation speed of multi-scenario analyses (mentioned here as multi-scenario calculation is not further discussed as calculation is similar to single-input set calculation).

Real-Time Visualization

During execution of a simulation run, a GUI to observe the system and model's current state has been implemented (Fig. 5.5, for illustrative purposes only, a description of the different elements is provided in the following). The pictured user interface is an example for a potential implementation and can be rapidly changed within Anylogic® if different reporting and visualization structures are needed.

On the top from left, the user interface provides general information on simulation parameters (e.g., simulated time horizon, progress, remaining simulation time),

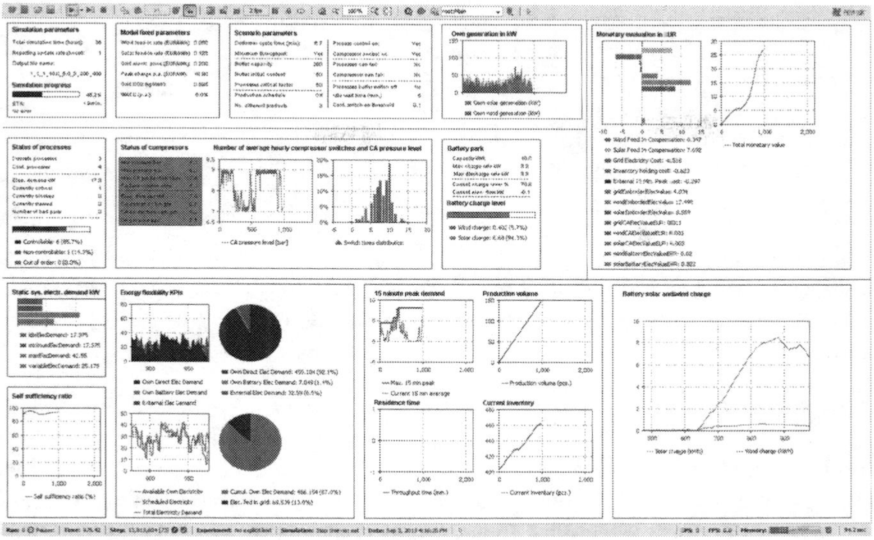

Fig. 5.5 Real-time indicator visualization cockpit: graphical user interface snapshot (for illustrative purposes only, a description of included elements is provided in this section)

fixed model parameters (e.g., feed-in rates, grid electricity price), scenario parameters (e.g., buffer capacity, if energy control is enabled, minimum throughput requirement) and a chart showing onsite generation time series. Within the second row, manufacturing process states (e.g., current electricity demand, number of blocked or starved processes) are summarized. Going further right, compressor states and charts, e.g., system pressure time series and switch-count histograms are shown. The battery park, including its parameters (e.g., capacity) and current state (e.g., charge level) is the next item. On the top right, a dynamically calculated chart shows the current monetary value generated by the control strategy, split by its components and drawn as a time series. The bottom row includes static electricity indexes to give an indication about the potential energy flexibility capability of the system. Further, a time series for self-sufficiency is generated. A set of time series and pie charts illustrates own, external, and battery electricity demand (time series and cumulative) as well as available electricity supply, scheduled electricity demand, and actual total electricity demand. The last two boxes (bottom right) highlight selected KPIs over time, e.g., 15-minute peak demand, production volume or solar and wind battery charge over time.

Figure 5.6 shows an example process chain visualized in the simulation prototype. Buffer and process parameters (e.g., maximum capacity, production rate) are shown as well as dynamic variables (e.g., fill level, scheduled production rate, fixed and variable electricity demand). In addition, color changes of frames and fill colors are used to indicate additional process statuses (e.g., blocked/starved, out-of-order, producing, critical).

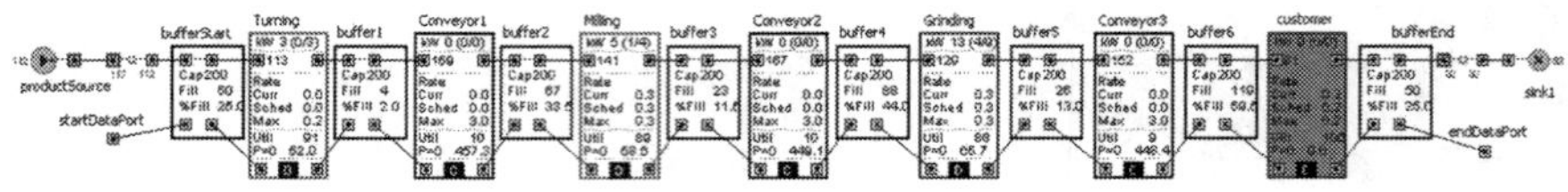

Fig. 5.6 Real-time production process visualization: graphical user interface snapshot (for illustrative purposes only, a description of included features is provided in this section)

Each agent of the simulation model and its parameters, variables and states can be observed as needed. In addition, Anylogic® offers the opportunity to generate plots and data-sets of every (dynamic) variable on-the-fly, offering additional visualization options if desired.

Output Data

Similar to input into the simulation model, output data is collected in an Excel file. All relevant simulation parameters (e.g., electricity control enabled, compressor park pressure bands, etc.) are recorded to allow traceability of simulation runs, followed by defined indicators (e.g., self-sufficiency ratio, monetary value). In addition, all validity checks (e.g., electricity supply-demand net balance, minimum and maximum CA system pressure) are included to control validity of a given simulation run followed by additional indicators and data-sets (e.g., embodied electricity by source and on a per-product basis, electricity demand time series).

Multi-scenario analysis is realized by assigning each output file a unique identifier composed of scenario-describing parameters. Separate files are required to avoid file access problems caused by multi-threading. After completion of an experiment, all scenario results are consolidated into a single file via a Visual Basic script. Additional charts and indicators can then be drawn to evaluate experiment outcomes. For onsite generated electricity, time series data for wind and solar generation is used.

5.4 Intermediate Conclusion

This chapter briefly outlined the prototypical implementation of the mathematical model described in Chap. 4 into a simulation prototype. Example model elements and agents are described, and their verification and behavior discussed.

In summary, a working, universally applicable Anylogic® prototype has been successfully implemented. The prototype is modularily built, allowing for fast adaption of scenarios and different application cases by simple drag-and-drop of elements. Further, the prototype does not rely on explicitly defined control periods or update rates, i.e., all value updates and control actions are event-triggered, ensuring high accuracy of the simulation prototype's calculations.

The next chapter describes example application cases for which the developed prototype has been applied to accomplish different objectives in the context of energy flexibility improvement of manufacturing systems.

References

Afazov, S. M. (2013). Modelling and simulation of manufacturing process chains. *CIRP Journal of Manufacturing Science and Technology*, *6*(1), 70–77. http://dx.doi.org/10.1016/j.cirpj.2012.10.005.

Balci, O. (1998a). Verification, validation, and accreditation. In D. J. Medeiros, E. F. Watson, J. Carson & M. S. Manivannan (Eds.), *Proceedings of the 1998 winter simulation conference*. Washington, DC.

Balci, O. (1998b). *Handbook of simulation, verification, validation and testing* (pp. 335–393). New York: Wiley.

Banerjee, S. (2014). *Mathematical modeling: Models analysis and applications*. Boca Raton: Taylor and Francis.

Banks, J. (1998). Principles of simulation. *Handbook of simulation: Principles, methodology, advances, applications, and practice* (pp. 3–30). New York: Wiley.

Chung, C. A. (2004). *Simulation modeling handbook: A practical approach*. Boca Raton: CRC Press.

Du, S., Xu, R., Huang, D., & Yao, X. (2015). Markov modeling and analysis of multi-stage manufacturing systems with remote quality information feedback. *Computers & Industrial Engineering*, *88*, 13–25. http://linkinghub.elsevier.com/retrieve/pii/S0360835215002715.

Fiedler, T., Ott, S., & Metz, D. (2007). Künstliche Neuronale Netze (KNN) zur Verbrauchsprognose im Strom- und Gasbereich. *Querschnitt*, *21*, 135–138.

Herrmann, C., Thiede, S., Kara, S., & Hesselbach, J. (2011). Energy oriented simulation of manufacturing systems - Concept and application. *CIRP Annals - Manufacturing Technology*, *60*(1), 45–48. http://linkinghub.elsevier.com/retrieve/pii/S0007850611001284.

Hesselbach, J. (2012). *Energie- und klimaeffiziente Produktion: Grundlagen, Leitlinien und Praxisbeispiele*. Wiesbaden: Springer Vieweg.

Kuhn, A., Reinhardt, A., & Wiendahl, H.-P. (1993). *Handbuch Simulationsanwendungen in Produktion und Logistik*. Wiesbaden: Springer Fachmedien.

Li, L., & Sun, Z. (2013). Dynamic energy control for energy efficiency improvement of sustainable manufacturing systems using Markov decision process. *IEEE Transactions on Systems, Man, and Cybernetics: Systems*, *43*(5), 1195–1205. http://ieeexplore.ieee.org/lpdocs/epic03/wrapper.htm?arnumber=6519950.

Li, L., Chang, Q., Ni, J., Xiao, G., & Biller, S. (2007). Bottleneck detection of manufacturing systems using data driven method. In *IEEE International Symposium on Assembly and Manufacturing* (pp. 76–81). Ann Arbor, MI. http://ieeexplore.ieee.org/lpdocs/epic03/wrapper.htm?arnumber=4288452.

Li, L., Sun, Z., & Tang, Z. (2012a). Real time electricity demand response for sustainable manufacturing systems: Challenges and a case study. In *8th IEEE International Conference on Automation, Science and Engineering* (pp. 353–357). Seoul. http://ieeexplore.ieee.org/lpdocs/epic03/wrapper.htm?arnumber=6386437.

Li, L., Sun, Z., Yang, H., & Gu, F. (2012b). Simulation-based energy efficiency improvement for sustainable manufacturing systems. *Proceedings of the ASME 2012 International Manufacturing Science and Engineering Conference MSEC2012* (pp. 1033–1039). IN: Notre Dame.

Liraviasl, K. K., Elmaraghy, H., Hanafy, M., & Samy, S. N. (2015). A framework for modelling reconfigurable manufacturing systems using hybridized discrete-event and agent-based simulation. *IFAC-PapersOnLine*, *48*(3), 1535–1540. http://dx.doi.org/10.1016/j.ifacol.2015.06.297.

Mattern, F., & Mehl, H. (1989). Diskrete Simulation-Prinzipien und Probleme der Effizienzsteigerung durch Parallelisierung. *Informatik-Spektrum*, *12*(4), 198–210.

Negahban, A., & Smith, J. S. (2014). Simulation for manufacturing system design and operation: Literature review and analysis. *Journal of Manufacturing Systems*, *33*(2), 241–261. http://dx.doi.org/10.1016/j.jmsy.2013.12.007.

ProModel. (2015). ProModel. Retrieved October 22, 2015, from https://www.promodel.com/Products/ProModel.

Rabe, M., Spieckermann, S., & Wenzel, S. (2008). *Verifikation und Validierung für die Simulation in Produktion und Logistik: Vorgehensmodelle und Techniken*. Heidelberg: Springer.

Schönemann, M., Herrmann, C., Greschke, P., & Thiede, S. (2015). Simulation of matrix-structured manufacturing systems. *Journal of Manufacturing Systems*, *37*(1), 104–112. http://dx.doi.org/10.1016/j.jmsy.2015.09.002.

Siemens. (2015). Logistics and Material Flow Simulation. Retrieved October 22, 2015, from http://www.plm.automation.siemens.com/en_us/products/tecnomatix/manufacturing-simulation/material-flow/index.shtml.

Sterman, J. D. (2000). *Business dynamics: Systems thinking and modeling for a complex world*. Boston: The McGraw-Hill Companies Inc.

Sun, Z., & Li, L. (2014). Potential capability estimation for real time electricity demand response of sustainable manufacturing systems using Markov decision process. *Journal of Cleaner Production*, *65*, 184–193. http://linkinghub.elsevier.com/retrieve/pii/S0959652613005738.

Tako, A. A., & Robinson, S. (2012). The application of discrete event simulation and system dynamics in the logistics and supply chain context. *Decision Support Systems*, *52*(4), 802–815. http://dx.doi.org/10.1016/j.dss.2011.11.015.

The Anylogic Company. (2015). Anylogic. Retrieved October 22, 2015, from http://www.anylogic.com/.

Thiede, S. (2012). *Energy efficiency in manufacturing systems*. Heidelberg: Springer.

Thiede, S., Seow, Y., Andersson, J., & Johansson, B. (2013). Environmental aspects in manufacturing system modelling and simulation-state of the art and research perspectives. *CIRP Journal of Manufacturing Science and Technology*, *6*(1), 78–87.

Yamin, H. Y., Shahidehpour, S. M., & Li, Z. (2004). Adaptive short-term electricity price forecasting using artificial neural networks in the restructured power markets. *International Journal of Electrical Power & Energy Systems*, *26*(8), 571–581.

Chapter 6
Example Application

Two case studies are used to demonstrate the applicability and effectiveness of the energy flexibility improvement concept and its implementation into a prototypical simulation tool. Two existing manufacturing systems are modeled. Data (parameters and structure) has been collected and the simulation prototype from Chap. 5 has been configured accordingly. The control approach's effectiveness is demonstrated by evaluation and comparison of selected indicators. Improvement targets are defined, parameters changed, and results evaluated. Both case studies show the successful application of the concept and provide examples for energy flexibility improvement options.

6.1 General Parameters and Structure

The following input parameters are held constant for all application cases and scenarios, unless otherwise noted:

- Grid electricity price: 0.197 EUR/kWh (company specific, example value from energy costs from TU Braunschweig, Germany in 2014 (Jaspers and Smyrek 2014)).
- Wind feed-in compensation: 0.089 EUR/kWh (value for onshore wind energy within first five years after commissioning according to German *Erneuerbare-Energien-Gesetz* (Bundesrepublik Deutschland 2014)).
- Solar feed-in compensation: 0.0879 EUR/kWh (according to German *Erneuerbare-Energien-Gesetz*, value from Bundesnetzagentur (2015), fixed feed-in rate for nonresidential building installation, commissioned after August 01, 2008, degression rate for September 2014).
- Peak demand cost: 131.04 EUR/kW p.a. (supply contract specific, average from two examples, Stadtwerke Rendsburg 2012, and Lechwerke 2015).

J. Beier, *Simulation Approach Towards Energy Flexible Manufacturing Systems*, Sustainable Production, Life Cycle Engineering and Management, DOI 10.1007/978-3-319-46639-2_6

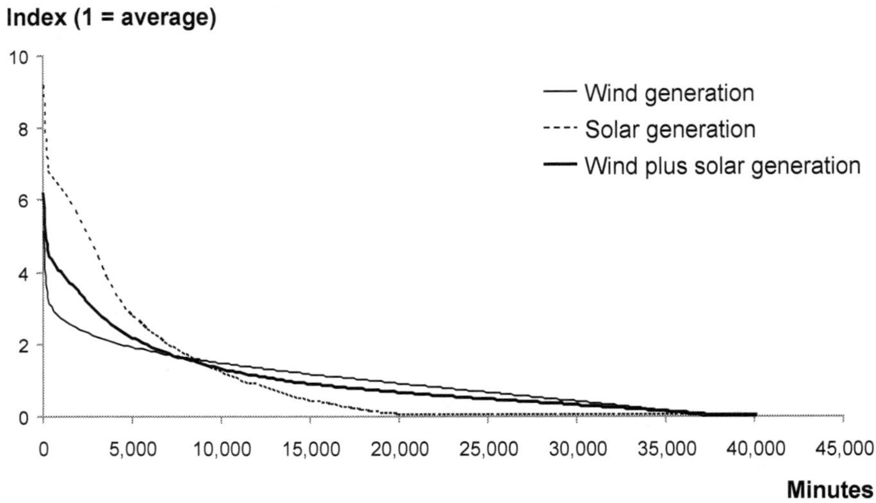

Fig. 6.1 Duration curves for wind, solar, and combined generation at Braunschweig, Germany, from September 3 to 31, 2013 (indexed to 1 for line-individual average, own illustration)

- Weighted average cost of capital: 6 % p.a. (company specific, example value).
- Carbon intensity per kWh grid electricity: 0.595 kg/kWh (Icha 2014).

VRE supply profiles have been obtained from generation facilities in Braunschweig, Germany. In particular, September 3 to 31, 2013 was chosen as sample period due to data availability. Recorded data profiles indicate that a sample period is used which exhibits both minute, hourly, and daily variations (compare also to Figs. 2.12 and 4.2). Respective duration curves for wind, solar, and combined generation can be found in Fig. 6.1. Wind output exhibits less extreme values and a longer, lower/steady output, while solar output has several heavy-output minutes (up to nine times higher than average) and more than half of the time with nearly zero output (at night). Data collection sample rate was one second (2,419,200 samples), which has been aggregated to one-minute averages to manage computability (40,320 samples).

6.2 Achieving an Autarkical Manufacturing Line

The first application example aims at demonstrating the concept's idea towards achieving an autarkical, i.e., electricity self-sufficient, manufacturing system.

6.2.1 Background and Hypotheses Formulation

The described case study is based on an experimental manufacturing lab at the Institute of Machine Tools and Production Technology (IWF) at TU Braunschweig,

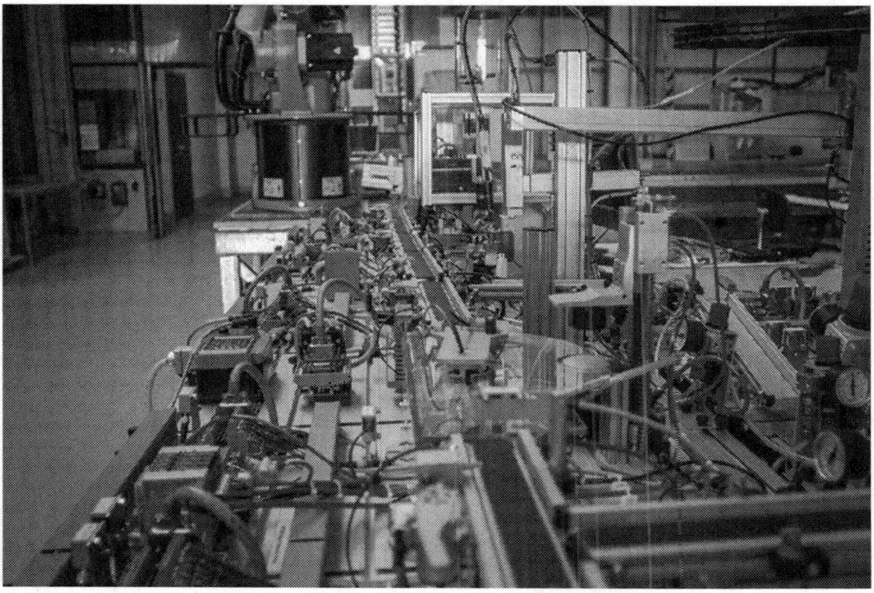

Fig. 6.2 Section of the experimental manufacturing lab at IWF, TU Braunschweig

Germany (Fig. 6.2). A sequential manufacturing line for teaching energy efficiency in manufacturing (among others) has been recently installed. Further, on-site electricity generation from wind and PV on the rooftop housing the manufacturing line is available, from which data has been collected as described before (cf. Sect. 6.1). In addition, EVs are used at the institute, which are also integrated into the application case study.

Referring to the application cycle provided in Fig. 4.23, the first step (define energy flexibility vision) has already been outlined: an autarkical manufacturing system enabled through VRE integration shall be achieved. The quantifiable goal (step two) is given by a self-sufficiency ratio of 100 %. Initial hypotheses to achieve this goal are (step three):

H1 Integration of available electric vehicles with grid-to-vehicle (G2V) and vehicle-to-grid (V2G) functionality (charging from and discharging into the local grid) can contribute towards integration of VRE if electricity generated from VRE is intermittently stored in EV batteries.

H2 EV integration is unlikely to enable full (100 %) self-sufficiency. Consequently, embodied energy storage through energy flexibility and (stationary) battery storage is investigated.

H3 Different VRE supply levels (amount of deployed capacity) and weighting between wind and solar capacities need to be considered to achieve an improved outcome from embodied energy and battery storage.

	Distribution A	Distribution B	Transport A	Assembling	Press	Switch	Industrial furnace	Transport B	CNC
Cycle time	8.0 s	2.5 s	4.0 s	10.0 s	8.0 s	6.0 s	17.0 s	4.5 s	72.0 s
Elec. idle [W]	960	990	1,000	1,400	920	1,160	4,430	1,080	4,060
Elec. prod. [W]	3,230	3,150	3,120	1,950	1,080	2,500	6,070	2,860	7,800
CA idle [Nm³/min]	0	0.54	0.20	0.20	0.89	0.16	0	0.17	0.17
CA prod. [Nm³/min]	0	1.43	0.57	2.81	3.70	0.17	0	0.46	0.20

Process Buffer

Fig. 6.3 Experimental manufacturing lab process structure and selected parameters (own illustration)

H4 Different battery configurations (power and energy values) need to be tested to understand if increased VRE utilization is enabled through improved energy flow (power) or improved energy storage.

H5 A combination of all above measures, also with energy efficiency options, can yield additional and favorable system configurations towards an autarkical manufacturing system.

H6 In order to understand the influence of different system utilization rates, changing target production rates (withdrawal rates from the last buffer) are evaluated.

Note that total production volume is held constant for all system setups and evaluation results, unless otherwise noted.[1] Within the next section, the system structure and parameters are described.

6.2.2 *General Structure and Parameterization*

Step four from the application cycle is determining the manufacturing system model structure, followed by evaluating parameter structure (step five). Figure 6.3 outlines the manufacturing line configuration from the experimental manufacturing lab.[2] The manufacturing line encompasses nine sequential process steps. Intermediate buffers are indicated between steps, with varying capacity to investigate different embodied energy storage options. A compressor park is connected to the system to supply CA, with an initial CA system volume ($CASV_{sys.}$) of 7 m^3 and three compressors with 6/4/2 Nm^3/min maximum supply and 4.2/2.8/1.4 kW production electricity demand, 60 s idle wait time and 10 % idle electricity demand (in percent of production electricity demand). In the base case, a customer demand rate (withdrawal from last buffer) of one product per 80 s is assumed, which is close to the maximum possible

[1]Minor deviations due to numerical methods and/or delayed adjustment of production state can be observed. However, no deviation was significant (within 10^{-1} percent range).

[2]Note that energy demand values have been multiplied by 100 to enable a better comparison to other system elements, e.g., battery and EV data.

Table 6.1 Electric vehicle parameters according to manufacturer's data

Name	Mia miAmore	Citroen C-Zero
Capacity [kWh]	8	16
Oper. range [km]	80	150
(Dis-)charge rate [kW]	2.667	2.667
Full (dis-)charge time [h]	3	6
Electricity demand [kWh/100 km]	10	10.67

throughput (defined by the process *CNC*, with the longest cycle time of 72 s). As noted above, simulation time is from September 3 to 31, 2013.

6.2.3 Hypothesis Testing

Steps six to eleven from the application cycle encompass building and validating the required manufacturing system model. Therefore, the prototype described in the previous chapter is configured according to outlined structure and parameters. The next step (twelve) is scenario definition to evaluate hypotheses.

Integration of Electric Vehicles

First, the effect of integrating electric vehicles into the manufacturing system and their contribution towards increased utilization of VRE is evaluated (cf. **H1**).[3] The connected EV fleet is assumed to consist of four vehicles of two different types. The first vehicle type is a Mia miAmore, the second type a Citroen C-Zero. Relevant vehicle type data can be found in Table 6.1. EVs are considered to behave similar to batteries modeled in Sect. 4.3.4: Battery/EV(s), with the exception that EVs and thus batteries might be temporarily unavailable. Further, vehicles and connected infrastructure are assumed to be (technically) capable of charging and discharging EVs from and into the local grid (see Fig. 6.4 for a picture of an EV charging station next to the building, where the manufacturing lab is located). Considered EV batteries are assumed to have a cycle lifetime ($Cyc_b^{80\,\%}$) of 1,200 cycles, a round-trip efficiency (η_b) of 90 % and a self-discharge rate (ξ_b) of 0.1 % per day. VRE supply was set to equal total system demand in a one-piece flow case.

Including EVs as intermediate electricity storage option has the advantage that batteries, which are already available, can be used to integrate VRE. However, vehicles are frequently removed and thus unavailable for charging and discharging. As such, the following scenarios assume that electricity from VRE is first used directly in the manufacturing system, then (if surplus exists) used to charge EVs and only

[3] This scenario and hypotheses are based on Beier et al. (2016a).

Fig. 6.4 Electric vehicle charging station and electric vehicle at IWF, TU Braunschweig

if both options are unavailable fed into the grid. If an EV is removed for driving, the current EV charge becomes unavailable. Driving distance determines how much charge is removed and thus also the amount of remaining electricity when the vehicle returns. Further, it is assumed that vehicle removal times are known upfront, and feeding electricity back to the manufacturing system is not allowed if remaining time for charging until removal is lower than required time to fully charge the EV's battery. The following scenarios are evaluated (REF refers to the reference scenario without battery and a one-piece flow production control):

- A stationary battery is set as a reference scenario, labeled BAT. The battery is assumed to have the same parameters as the four EVs ((dis-)charge rate and capacity).
- Different vehicle utilization schedules are tested (Fig. 6.5). The first schedule (SCH1) indicates a heavy-use scenario, where all vehicles are unavailable throughout the day. SCH2 is a high-frequent scenario, with low driving distances. A mixed schedule scenario, based on actual/observed utilization patterns, is modeled in SCH3.
- The effect of additional EVs and thus battery storage is evaluated by doubling the amount of vehicles. All scenarios remain the same, scenarios with four vehicles are labeled *a* (e.g., SCH1a) and scenarios with eight vehicles are labeled *b* (e.g., SCH1b).

Figure 6.6 summarizes the results for all nine scenarios. Overall, an increase in VRE utilization can be achieved through EV integration. Compared to the refer-

Scena-rio	Vehicle	Time (24h)	Mon	Tue	Wed	Thu	Fri	Charge removed
SCH1	All	08:00-16:00	✓	✓	✓	✓	✓	90%
SCH2	All	08:00-10:00	✓	✓	✓	✓	✓	20%
		11:00-12:00	✓	✓	✓	✓	✓	30%
		13:00-14:00	✓	✓	✓	✓		20%
		15:00-16:00	✓	✓	✓	✓		30%
SCH3	M1	10:00-13:00		✓	✓	✓	✓	50%
		15:00-18:00		✓	✓	✓		40%
	M2	10:00-11:00	✓	✓	✓	✓	✓	10%
		13:00-15:00		✓		✓		30%
		18:00-08:00	✓	✓		✓		40%
	C1	08:00-16:00	✓	✓	✓	✓		80%
	C2	09:00-12:00	✓	✓	✓	✓	✓	30%
		14:00-15:00	✓	✓		✓		20%

Fig. 6.5 Vehicle operating schedule scenarios (M1/M2: Mia miAmore 1/2, C1/C2: Citroen C-Zero 1/2, own illustration)

ence scenario, EV integration (four vehicles) achieves an increase in self-sufficiency (SSR_T) up to 68 % (from 62 % in the base case). A stationary battery would add another 2 %, which indicates that even with frequent removal/utilization of EVs, a substantial self-sufficiency increase can be achieved. Doubling the amount of vehicles further increases self-sufficiency. However, the relative improvement is lower compared to the first scenario (i.e., 8 % increase from REF to BATa, but only 6 % increase from BATa to BATb), indicating that doubling energy storage capacity does not result in doubling self-sufficiency improvement due to VRE supply dynamics (i.e., high PV generation which exceeds charge rate of batteries). Further, if EVs are removed (driving-schedule scenarios), relative improvement of eight vehicles compared to four vehicles is also lower compared to four vehicles and no battery storage.

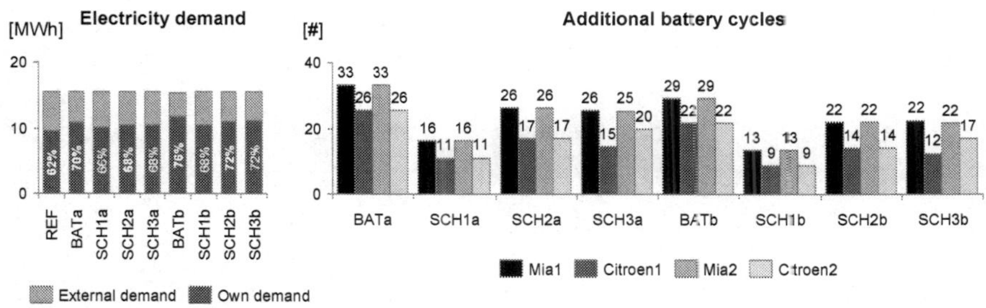

Fig. 6.6 Results for EV integration scenarios (own illustration)

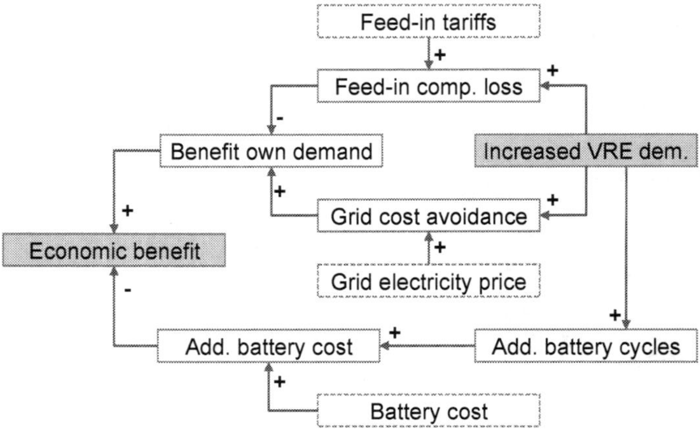

Fig. 6.7 Dependence between increased VRE demand through battery storage and potential economic benefit/loss as a function of feed-in tariffs, grid electricity price and battery cost (own illustration)

Using EVs for intermediate electricity storage has the disadvantage that batteries are stressed more and thus lifetime is reduced. The right graph in Fig. 6.6 shows additional battery cycles ($Cyc_{b,T}$, no driving-related cycles included) imposed on batteries due to VRE storage and discharge. Depending on the specific scenario, up to 33 additional cycles (BATa) are required within the 28 days simulation time. This indicates, assuming a cycle lifetime of 1,200 cycles, that the EV battery's capacity is reduced to 80 % after approximately 40 months, only from VRE storage for use in the manufacturing system. As capacity of EV batteries is essential to ensure a long driving distance, reduced capacity is a large disadvantage. Further, assuming grid electricity costs and feed-in tariffs as outlined in the beginning of this chapter, slightly above 100 Euros are gained in the BATa scenario. Considering that four EV batteries might require replacement after 40 months, utilizing EV batteries is a rather uneconomic option for VRE integration.

Figure 6.7 illustrates the dynamics between increased VRE utilization and economic benefit/loss as a function of feed-in tariffs, grid electricity price and battery cost. Depending on the specific relation between parameters and VRE demand, including EV batteries can be economically beneficial. Specifically, if parameters, especially battery cost, change, a previously uneconomic scenario can become favorable.

After EV integration has been evaluated, embodied energy storage and stationary batteries with different capacities are tested.

Embodied Energy and Battery Storage

As outlined in the concept, a central strategy for VRE integration is utilizing embodied energy storage in (intermediate) products. As such, different intermediate storage capacities between processes are combined with an enacted central energy control.

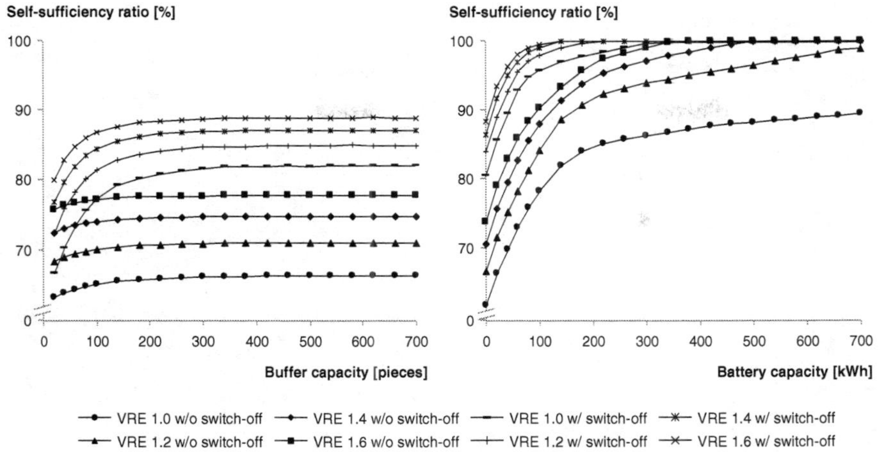

Fig. 6.8 Self-sufficiency for different intermediate buffer capacities, battery capacities, VRE supply scenarios and energy efficiency control actions (w/o: without, w/: with, VRE: variable renewable electricity, own illustration)

Further, in order to compare embodied energy storage with (electrical) energy storage, different battery sizes are also evaluated (cf. **H2**). In addition, different VRE supply levels are combined with embodied energy storage and battery storage to investigate the influence of VRE (over-)supply on the potential for an autarkical manufacturing system (cf. **H3**). Further, to be able to compare all previous scenarios and hypotheses, self-switch-off for processes is included in above experiments (cf. **H5**).[4]

Results for described experiments can be found in Fig. 6.8. A VRE share of 1.2 indicates that total VRE supply is 20 % higher than the manufacturing system's demand in the base case (one-piece flow). Two different energy efficiency control actions are indicated by "w/o switch-off," which leaves all processes on (idling), and "w/ switch-off," which indicates that processes are switched off if 1 min idle wait time has passed. First, pictured in the left graph, self-sufficiency for different intermediate buffer capacity sizes and VRE share is illustrated. Self-sufficiency can be increased by deploying additional intermediate product storage. However, the effect is lower if (a) already large intermediate product storage exists (moving right on the horizontal axis) and (b) the level of VRE oversupply is increasing (comparing 1.0 VRE to 1.2 VRE and 1.4 VRE to 1.6 VRE) and (c) both effects are combined (increasing buffer capacity if VRE is already high has a smaller impact than increasing buffer when VRE supply is low). Nonetheless, enacting energy efficiency control significantly improves self-sufficiency. Processes are switched off if idling and only on if sufficient VRE supply is available. Increasing intermediate product storage has a substantial impact as processes can be more flexibly switched off. Figure 6.9 illustrates external and own electricity demand (absolute values) in relation to buffer capacity for a 1.2

[4]Results from this subsection are also discussed in Beier et al. (2016b).

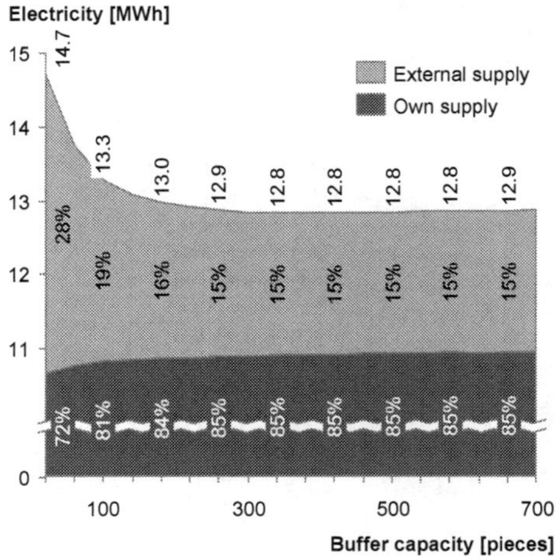

Fig. 6.9 External and on-site VRE demand in relation to intermediate buffer capacity with 1.2 VRE supply factor and process switch off (own illustration)

VRE supply scenario and with process switch off. Additional buffer capacities do not only result in an increased energy flexibility and thus demand of on-site VRE, but also significantly reduce total electricity demand. Further, due to the central energy flexibility control, nearly only external demand is reduced.

A fully (100 %) autarkical operation of the manufacturing system as envisioned has not been feasible in any previously discussed scenario. Therefore, the inclusion of electrical battery storage is evaluated. The right graph from Fig. 6.8 shows self-sufficiency over different battery capacities, which are included in the system (note that (dis-)charge time has been held constant at three hours). Further, the VRE supply cases labeled without switch off indicate that no energy control actions were enacted, i.e., a pure one-piece flow production control was present. Cases with switch off represent 200 pieces intermediate buffer capacity with energy flexibility control and process switch off.

Overall, including battery storage enables a fully autarkical operation of the manufacturing system. However, significant oversupply of VRE is still required: for the VRE cases without switch off, an autarkical operation within the evaluated battery range is only feasible above 1.2 times VRE oversupply, i.e., for the 1.4 case, a battery of about 540 kWh is required to ensure self-sufficiency. With switch off, total energy demand is reduced (as indicated in Fig. 6.9), which also results in increased oversupply if VRE supply is held constant (comparing the two cases with and without switch off). As a result, battery capacity requirements are significantly reduced, and a full autarkical operation is already feasible with a 200 kWh battery and 1.2 VRE oversupply. Taking discussed findings from buffer and battery capacity variations into account, wind and solar share variations are tested within the following.

Wind and Solar Share Variation

Figure 6.10 summarizes self-sufficiency for different wind share values and battery capacities and three hours charge and discharge time. VRE oversupply has been held constant at 1.2 times, and energy flexibility control with switch off and 200 pieces intermediate buffer storage was enabled. Total on-site generated supply was held constant, but the contribution and thus magnitude of time series from solar and wind generation was altered, i.e., a wind share of 70 % implies that the remaining 30 % of total on-site supply is generated from solar energy.

Overall, varying wind share and battery capacity has a strongly nonlinear and also non-monotonically behavior in the direction of different wind shares, while, as already observed, a monotonically increasing behavior in the direction of increasing battery capacity exists. Further, the maximum self-sufficiency line as a function of battery capacity is not linear and shifting from high wind share values to low wind share values and back to high wind share values (e.g., max. self-sufficiency: 80 % wind at 40 kWh, 70 % wind at 80 kWh, 60 % at 180 kWh, 50 % at 210 kWh, 60 % at 240 kWh). In general, a wind share value of above 50 % appears to be favorable, while the specific optimum (for the evaluated configuration) is dependent on installed battery size. Further, a very high wind or solar share (and thus low share of the other supply source) is less favorable for increasing self-sufficiency. A mixture of both sources appears to complement each other under enacted energy flexibility control and installed battery storage. Further, choosing the right wind/solar share can enable full autarkical operation, i.e., with a battery capacity of 250 kWh, 60 % wind share enables 100 % self-sufficiency, while only 20 % would not allow a self-sufficient operation. However, a battery is required in any wind/solar combination for full autarkical operation. As such, different battery parameters are tested next.

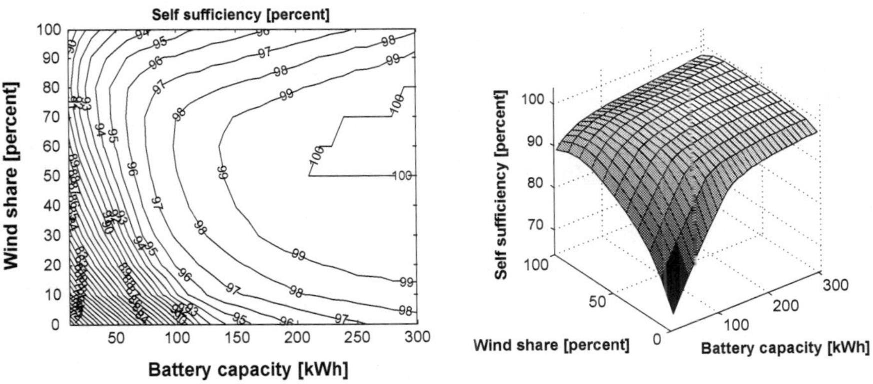

Fig. 6.10 Self-sufficiency ratios for different wind share and battery capacity values (note that discrete values are caused by chosen experiment resolution (ten unit steps in each direction), own illustration)

Battery Power and Capacity Sizing

In order to understand the influence of different battery characteristics, an experiment is executed which aims at determining the impact of different battery energy and power levels (cf. **H4**). Determining a power-energy combination which is favorable can be used to determine suitable battery technologies (cf. Sect. 2.3.1: Decentralized, small-scale energy storage).

Figure 6.11 summarizes the results for different battery energy (capacity) and power (charge time) values. Achieving a fully autarkical operation is only feasible with a battery capacity of around 210 kWh (note that step length between scenarios is 10 kWh). Further, (dis-)charge time has a negligible influence when an autarkical operation is pursued and has only notable influence for lower battery capacity values. For smaller capacity values, self-sufficiency is reduced if (dis-)charge rates are lowered (keeping capacity constant). High energy peaks (charge and discharge) cannot be accommodated if energy flow rates are too low.

In summary, battery capacity is (again) the clear main factor for achieving an autarkical operation, while (dis-)charge rates have a negligible influence. This finding can be used to chose a specific battery technology, i.e., by prioritizing economic and environmental factors rather than allowable energy flow rates. For the purpose of this application case, (dis-)charge rate is kept at three hours (unless otherwise noted) to allow better comparison between experiments.

Further Investigations on Embodied Energy Storage and Customer Cycle Time

After several scenarios on battery size, VRE share and process switch off have been evaluated, another set of experiments is executed to determine if battery capacity can be further reduced by improving embodied energy storage in intermediate products.

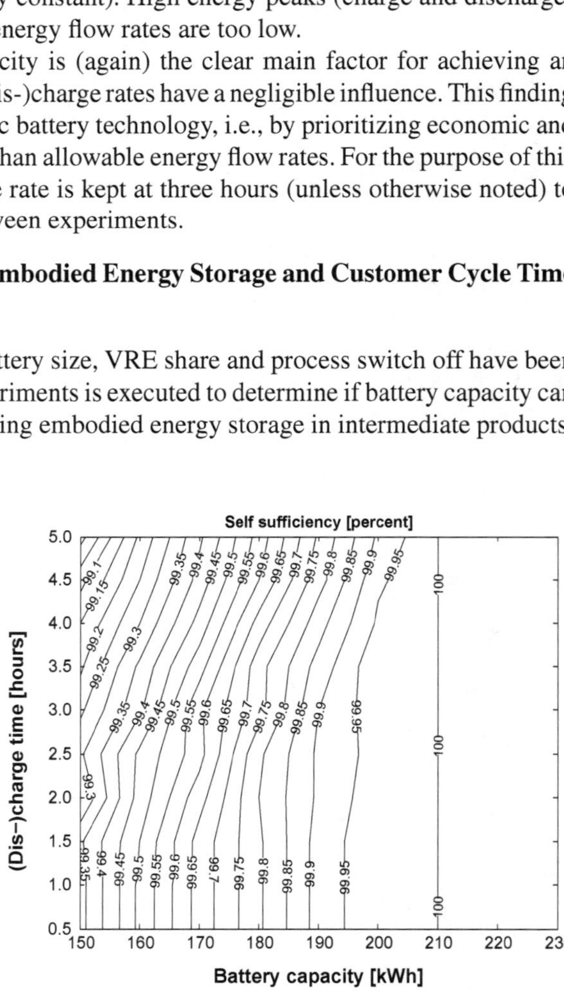

Fig. 6.11 Self-sufficiency ratios for different battery capacity and charge time values (own illustration)

As the last process, CNC, has the longest cycle time, not only different buffer values between all processes, but especially before and after CNC, are tested.

Figure 6.12 illustrates results for six different indicators under varying buffer capacity values between all processes (vertical axis) and separately before the CNC process (horizontal axis). Energy flexibility control and process self-switch-off were enabled, a battery capacity of 300 kWh (three hours (dis-)charge time) installed, VRE supply set at 1.2 times demand with 50 % wind and solar share.

Increased buffer space before CNC is expected to decouple CNC and thus improve energy flexibility. As the contour plot for self-sufficiency from Fig. 6.12 shows, pre-CNC buffer capacity can decrease buffer capacity requirements for remaining processes. For example, with a buffer capacity of 30 pieces between all processes and 330 pieces pre-CNC, 100 % self-sufficiency is enabled (scenario combination denoted as (330, 30)). For a (30, 30) case where all capacities are equal, only 96.8 % self-sufficiency is achieved. Further, with only 30 pieces pre-CNC storage, full autarkical operation is enabled at (30, 130), and keeping all buffers the same size, the scenario (110, 110) is the first case where full self-sufficiency is enabled. Aside from the trade-off between different buffer space locations, another conclusion can be drawn: for a given battery and VRE supply scenario, altering intermediate product storage and thus energy flexibility can significantly improve self-sufficiency towards an autarkical operation, and thus reduce battery size or other energy storage requirements.

The remaining graphs from Fig. 6.12 illustrate the multi-indicator evaluation for experiments. As noted before, both scenarios (330, 30) and (110, 110) achieve an autarkical operation. Total electricity demand is very similar for both cases (12,893 kWh and 12,899 kWh, respectively). However, average inventory is significantly lower for the pre-CNC scenario (438 pieces) than for the equal-buffer-size case (776 pieces), which translates also to different system residence times (578 min vs. 1,020 min). Considering monetary value (for simplicity, a constant machine hour rate of 60 EUR/h was assumed for all processes and no initial product value), the pre-CNC case achieves a value of 421 Euro, while the equal-buffer case achieves 405 Euros. Therefore, at least for the included items in the monetary evaluation, the pre-CNC case is slightly more beneficial, partly due to lower inventory and thus working capital which needs to be financed (this effect is increased with higher value products, i.e., an initially high product value). The last graph (lower right) shows CO_2 emissions per product, which exhibit a similar pattern than self-sufficiency (and are zero for the two discussed cases as 100 % RE input is achieved). In summary, the given example illustrates that multi-scenario/-experiment evaluation and different indicators are needed to chose a desired operating point of an (energy flexible) manufacturing system.

The last experiment discussed within this application case evaluates the impact of different customer demand rates and thus utilization of the system under multi-indicator evaluation (cf. **H6**). Lower utilization should increase self-sufficiency as scheduling flexibility exists. However, energy demand per product might be increased if switch on/off and idle electricity demand is distributed among less products. Further, the impact of different buffer sizes in relation to customer demand rates is

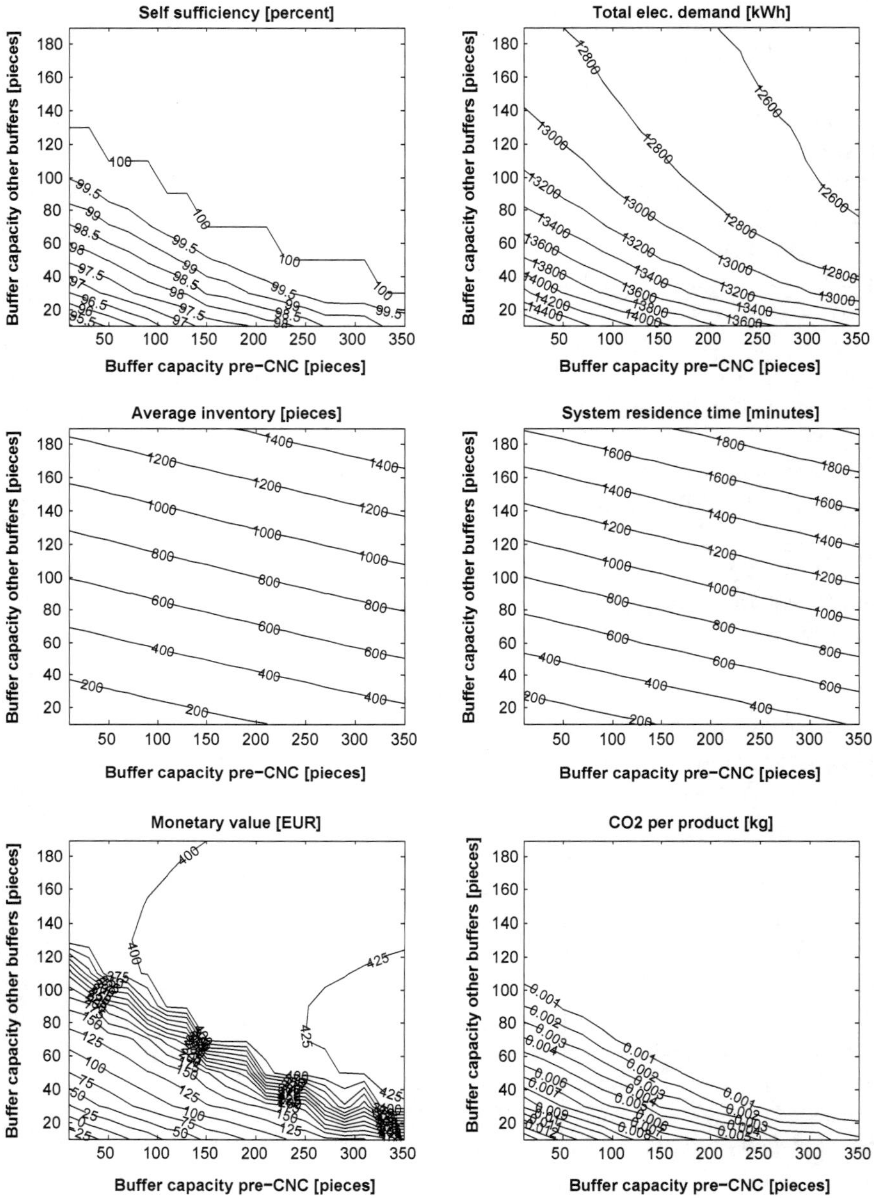

Fig. 6.12 Six indicator values for varying buffer capacities before CNC process and for all remaining processes (own illustration)

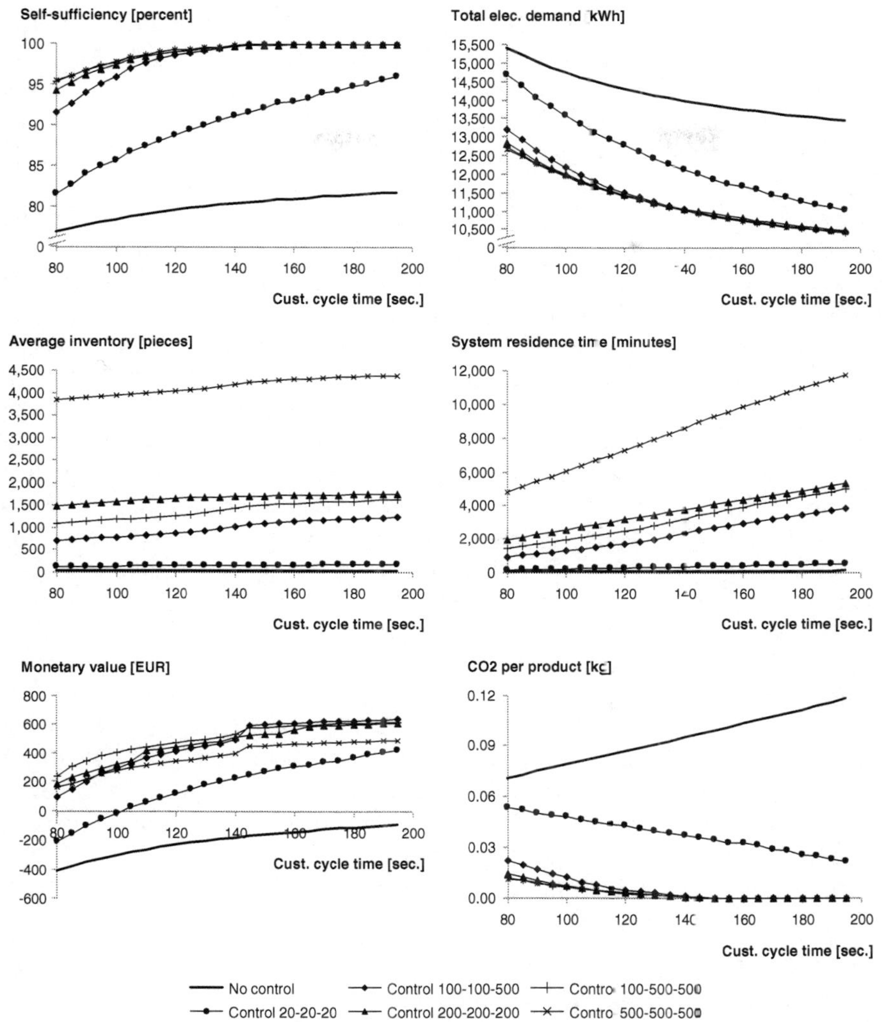

Fig. 6.13 Six indicator values for varying customer cycle times and intermediate buffer capacity scenarios (first value referring to all process buffers, second to pre-CNC and third value to post-CNC buffer, own illustration)

investigated to determine dependencies. Figure 6.13 provides results for the same indicators as discussed for the pre-CNC case and for customer cycle times between 80 and 195 s, indicating a utilization reduction of approx. 2.5 times (a battery with 50 kWh, 3 hour (dis-)charge time is included). Further, buffer capacity is varied for all buffers, pre-CNC and post-CNC buffer (similar notation to above with three values, e.g., (20, 20, 20)).

First, the no-control (one-piece flow) strategy achieves inferior performance for all indicators except inventory and residence time due to low target buffer levels.

Further, total electricity demand is nonlinearly decreasing due to fixed idle electricity demand, while CO_2 per product is linearly increasing due to the same fact (total values are nonlinearly increasing due to lower production volume for lower customer demand rates). Further, the (20, 20, 20) case illustrates that self-sufficiency and energy flexibility can be strongly increased with lower customer demand rates and thus throughput requirements, indicating a higher energy flexibility due to lower utilization. The other shown buffer-variation cases illustrate that further decoupling of the CNC process can also be beneficial for self-sufficiency and additional indicators, while an additional increase in all buffer capacities is only recommendable if CNC is not decoupled. Most notably, the (100, 500, 500) case achieves similar self-sufficiency values than the (500, 500, 500) case, with substantially lower inventory and residence time. In addition, full autarkical operation is supported by low utilization rates and feasible for several buffer capacity combinations at decreasing utilization rates (compared to the initial case of 80 s demand rate, which is close to maximum possible throughput). In summary, these experiments illustrate that decreasing utilization and thus throughput requirements are beneficial for autarkical operation of the system and thus energy flexibility. Total electricity demand is reduced, but nonlinear, indicating higher electricity per product as idle electricity demand is distributed among less products. This is in line with CO_2 per product, which is reduced, but only as a result of increased VRE demand (increased VRE share, but higher total electricity demand per product). Another negative impact can be observed for average inventory levels and system residence time. Monetary value improves with longer customer cycle times, however, the concept's definition of monetary value excludes certain effects of lower utilization rates, such as reduced sales and thus lower total contribution margin. As such, additional effects of lower production volume need to be carefully quantified. Reducing production volume is (potentially) only beneficial if additional products cannot be sold (e.g., as an effect of reduced market demand and thus manufacturing overcapacity).

6.2.4 Case Study Summary and Conclusion

Discussed case study illustrates that a combination of embodied energy storage, energy flexibility control, energy efficiency measures, and battery storage can be used to enable an autarkical operation of an example manufacturing system. Table 6.2 qualitatively summarizes selected scenario changes (influencing factors) on six KPIs. Although the table aims at providing a qualitative summary only (detailed values and sensitivity analysis cannot be substituted by indicating a single direction of influence) and only for the specific use case, it reflects that a holistic evaluation of multiple measures and their impact on relevant indicators is required to select a suitable strategy. Several measures improve energy flexibility to a different extent, but they might also negatively affect other KPIs. Depending on individual (energy flexibility planner's) preferences, one or more strategies can be chosen.

Table 6.2 Qualitative influence of scenario changes (left) on selected indicators (top, note that increasing product storage and longer customer cycle time is combined with process self-switch-off)

	Self-sufficiency	Total elec. demand	Avg. inventory	System residence time	Monetary value	CO_2 per product
Process switch off	↑	↘	↑	↑	↑	↓
Balanced wind/solar share	↗	↔	↔	↔	↔	↘
Incr. battery capacity	↑	↔	↔	↔	↗	↓
Incr. battery charge rate	↗	↔	↔	↔	↔	↘
Incr. product storage	↗	↘	↑	↑	↗	↘
Longer cust. cycle time	↑	↘	↗	↗	↗	↘

Although an autarkical operation can be achieved by deploying sufficient battery storage and/or VRE supply, above measures can reduce required capacities significantly. Considering that both battery and VRE capacities are costly and, especially for the case of batteries, potentially environmentally unfriendly, utilizing embodied energy storage appears to be a beneficial strategy. However, as discussed before, additional space and equipment, both for energy control actions and product storage, are required. Further, operational indicators such as system residence time and average inventory are less favorable under embodied energy storage.

Figure 6.14 shows an example time series for manufacturing system electricity demand and own supply, both for a no-control and a central energy flexibility control case. Further, the difference between supply and demand is given (a positive delta indicates grid supply, a negative delta grid feed). Within the first 20 min, demand and supply are relatively balanced, also without central energy control. However, own supply decreases thereafter, resulting in increasing grid dependency without central energy control. With central energy control, the delta between own supply and demand is significantly lower (lower graphs of Fig. 6.14), resulting in reduced grid dependence. Total absolute deviation between supply and demand is 19.13 kWh for the no-control scenario and 7.68 kWh for the control scenario. The central energy control logic successfully achieves a better fit of demand with on-site supply.

Additional hypotheses that emerged during evaluation of system behavior can be tested and interpreted in a similar way to above hypotheses and experiments. These include, for example, variation in CA storage tank size (see also Beier et al. 2015 for an example application of proposed concept towards flexible CA production and

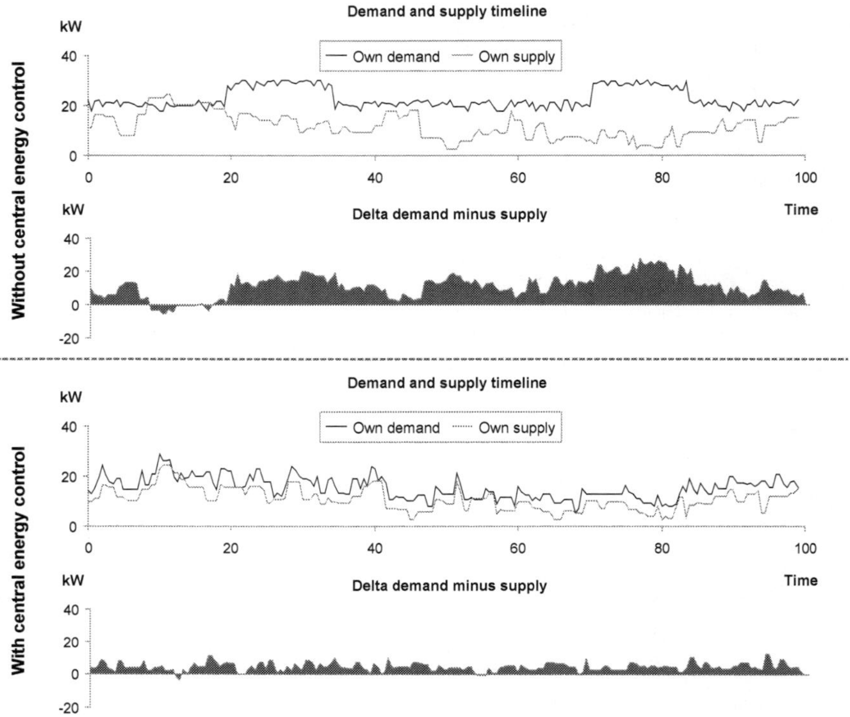

Fig. 6.14 Example time series for electricity supply and demand from the experimental manufacturing lab case study with and without central electricity control (time refers to minutes, sample period is from minute 1,200 to 1,300 (September 3, 2013 20:00 h to 21:40 h), own illustration)

storage), or selectively altering intermediate buffer storage size (e.g., only increasing product storage before CNC to decouple CNC from the remaining process chain as the process' cycle time is significantly longer than remaining processes' cycle time).

6.3 Aluminum Die Casting Case Study

The second example application case study presented here aims at demonstrating the concept's approach to another manufacturing process chain, an aluminum die casting line. Rather than focusing on different hypotheses testing towards energy flexibility goals, a focus is set on demonstrating additional (model) features from the concept and to discuss influencing factors for energy flexibility, including challenges. Additional background on the case study can be found in Thiede (2012) and Heinemann (2016), where the process chain has been used as an application example for energy efficiency-related simulation and in the context of (cross-company) energy and resource efficiency improvement in aluminum die casting.

6.3.1 Background and Process Structure

Figure 6.15 illustrates the aluminum die casting process chain. Several conceptual topics are addressed in addition to the first case study:

- Diverting and combining flows are included.
- Die casting machines (DCMs) exhibit a post-production time, during which different energy (electricity/CA) demand values are observable.
- Buffer capacity between DCMs and saws is very low (three pieces), thus material flow is closely linked.
- Different batch sizes are introduced, i.e., the amount of products that are processed simultaneously/at once.

Both post-production time as additional (mandatory) process state as well as batch sizes (larger than the original single-piece processing) require minor adjustments to the original model. Post-production is a mandatory wait time after production state with individual energy demand, while batch sizes larger than one are realized by postponing the production start until the required batch size is available to the process. Further, a new batch can only be started if the previous batch has been fully discharged into the outgoing buffer. For additional details on the modeled process chain see also Thiede (2012).

VRE supply was set to an equal supply and wind share, total cumulative supply is 1.1 times demand in a one-piece flow case (number of products in a batch for batch-type processes).

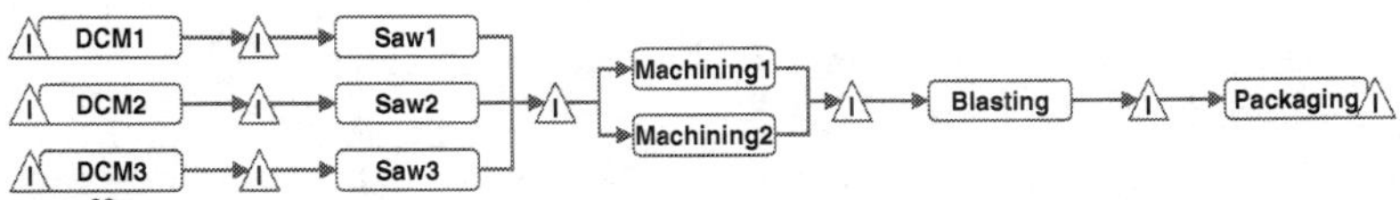

Cycle time/ buffer cap.	69 s (81 s)	3 pcs.	60 s	variable	270.0 s	variable	55.0 s	variable	30.0 s
Batch size [pcs.]	1		1		2		10		1
Elec. idle [kW]	187.10		2.70		1.33		51.60		0
Elec. prod. [kW]	219.40 (221.10)		7.70		13.30		74.30		1.00
CA idle [Nm³/min]	0		0		0		0		0
CA prod. [Nm³/min]	3.33 (0)		0.50		1.43		0.83		1.67

Process Buffer

Fig. 6.15 Aluminum die casting process chain and main parameters (note that values in parenthesis are post-production values; DCM: die casting machine, own illustration)

6.3.2 *Experiments and Results*

As mentioned before, a focus is set on highlighting additional conceptual features and discussing energy flexibility potential and influences rather than testing a dedicated set of hypotheses. Therefore, the initial (and general) hypothesis is to investigate the energy flexibility potential of the process chain and the influence of energy flexibility control (compressors and processes). The objective is to improve demand of on-site generated VRE, while investigating the influence on additional operational indicators under stochastic system behavior.

Experiments

Three different experiments are designed and aim at drawing conclusions on the following system setups:

- Stochastic failure of machines is introduced. All processes are configured with a MTBF of 270 s and a MTTR of 30 s (resembling 90 % availability). Due to tight inventory levels/buffer capacity and batching, an impact on overall production volume is likely. Other indicator results will also be stochastic. A no-control case is introduced (one-piece flow with batching) and 120 iterations/simulation runs evaluated.
- Low buffer capacity values are likely to limit energy flexibility in product embodied energy storage. Therefore, energy flexibility using flexible CA generation is tested. Four compressors with 6/4/3/2 Nm^3/min production rate and 42/28/21/14 kW production energy demand are included in the system, connected to a 70 m^3 storage tank and controlled via energy flexibility control. All other parameters are held constant compared to the previous experiment, including random failures of machines.
- In order to compare results to embodied energy storage, a third experiment enacts compressor and process energy flexibility control, leaving all other parameters constant.

Results

Figure 6.16 shows result histograms for described three experiments and five indicators. First, comparing different production volume values, production volume is not constant and distributed randomly. For the no-control case, median production volume is 24,402 pieces, with 352 pieces range between the first and third quartile. Considering a cycle time of 90 s, maximum throughput without failures would be 26,880 pieces, and thus 90.8 % of maximum throughput is achieved (note that without a mandatory customer withdrawal rate, the two machining processes have the longest combined cycle time of 67.5 s per product). With compressor control only, production volume is, as expected, within the same range. However, with process control, production volume is reduced to 21,567 pieces (minus 11.6 % compared to no-control). Even though bottleneck detection and bottleneck dependent control

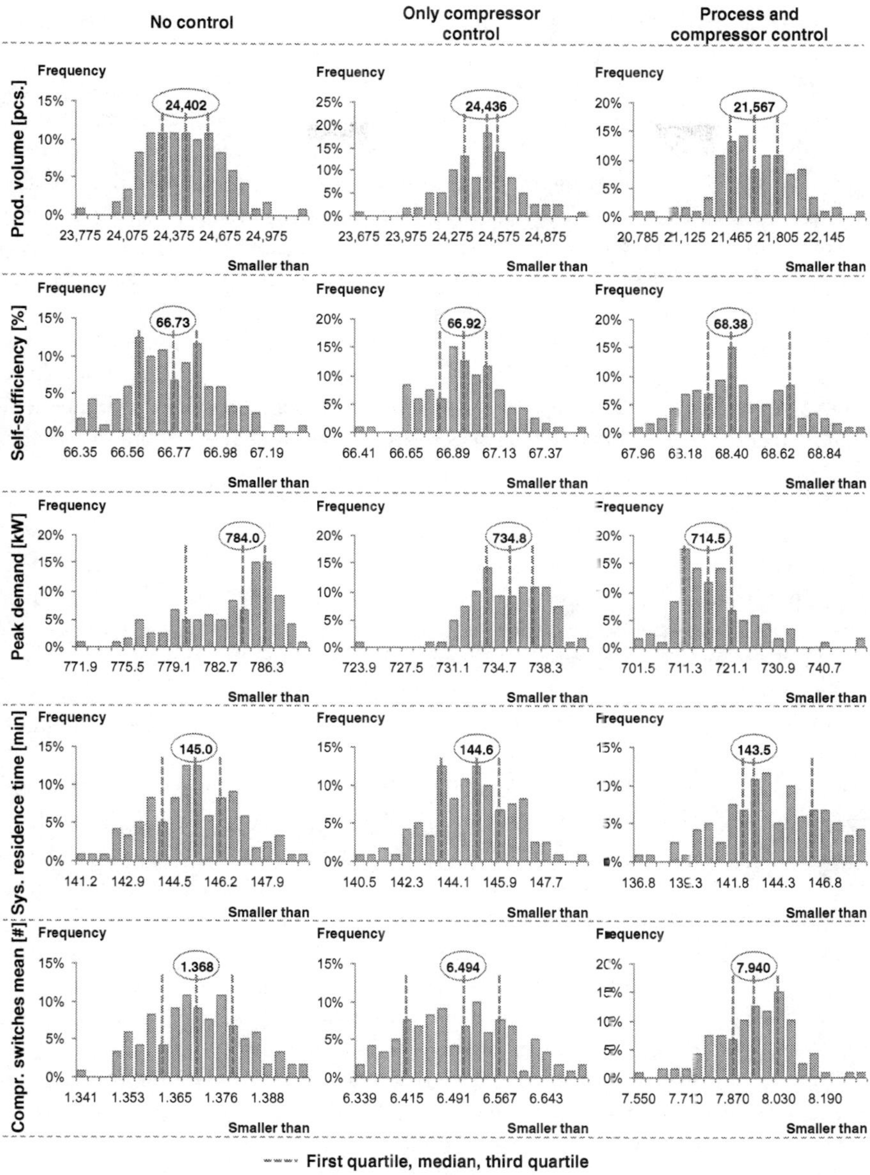

Fig. 6.16 Result histograms for five indicators and three experiments for the aluminum die casting application case (120 simulation runs/results per diagram, own illustration)

is enacted, low intermediate buffer capacities combined with random failures and large batches in relation to buffer capacity cause processes to not be able to react fast enough to bottleneck-critical scheduling (bottleneck processes might be blocked or starved more often if energy control sets processes to idle). Variance/difference between first and third quartile remains similar (363 pieces).

Considering impact on self-sufficiency, compressor control alone achieves only a negligible median increase (statistic significance testing is omitted as not relevant for further conclusions), while combined compressor and process control achieves approx. a 1.5 % increase, which is, considering values from the previous application case, also minor. 15-minute peak demand is on average reduced through central energy control. System residence time is relatively constant between experiments and narrowly distributed as a result of small intermediate product storage capacities. Average compressor switches per hour are around 1.35 without energy control and significantly increased as a result of energy control actions. Maximum compressor switches for all compressors (median of all 120 evaluations) is two switches maximum without control, 22 switches with compressor control and 24 switches with combined control (not illustrated here). Specific distribution and maximum switches per compressor needs to be closely monitored to not exceed maximum allowed switches.

In summary, this application case study demonstrates the impact of parallel flows, mandatory production sequences (post-production), batches, and stochastic variables within proposed energy flexibility concept. Obtained indicators exhibit stochastic behavior and require advanced simulation and evaluation (obtaining a sufficiently large sample set). As such, required simulation time for an experiment/scenario significantly increases. However, the application case demonstrates that the concept is applicable under mentioned process and system characteristics (stochastic failures, parallel flows).

Considering energy flexibility potential of the die casting process chain, integrating on-site VRE appears to be challenging. Embodied energy storage in products has limited potential as intermediate product storage capacities are low, processes exhibit a long cycle time and/or large batch size. In addition, the processes with (by far) largest electric power demand are the three DCMs, which are, in turn, exhibiting a mandatory post-production time and are closely linked to the three saws with only 3 pieces intermediate storage. In addition, stochastic failures combined with energy control, linked processes and large and very different batch sizes between processes significantly reduce throughput under energy control actions.

For the case of CA energy storage, potential is very limited as production machines only demand little CA compared to electricity. Median average CA demand without control for the 120 simulation runs is 4.2 Nm^3/min., which equates to approx. 29.4 kW electricity demand from compressors. In turn, total system (processes and compressors) average electricity demand is 632.7 kW. Thus, process electricity demand is more than 20 times larger than compressor demand, limiting potential for energy flexible CA production. Even with a three times larger CA storage tank (210 m^3), median self-sufficiency increased only to 67.15 % (from initially 66.92 %). For another example focusing on energy flexibility enabled by CA storage, including tank size variation, see Beier et al. (2015).

In order to improve energy flexibility of the die casting process chain, the effect of additional intermediate product storage could be investigated. However, as discussed, the DCMs require decoupling most as they demand the largest share of overall power. Decoupling is not feasible as limited storage is available between DCMs and saws in the existing layout. Another option is to always run the saw immediately if a part is available to combinedly decouple DCMs and saws. However, this strategy results in long mandatory energy demand patterns if DCMs are set into production mode and might thus also negatively affect energy flexibility (while also random failures can prevent direct production start of saws). Switching electricity demand to CA demand (fuel switching) has also limited potential as CA cannot be used for heating purposes. An additional strategy would be to attach a stationary battery to the system, which should increase VRE utilization. However, similar to the first application case study, a battery requires significant additional investments and increases (lifetime) environmental burden of VRE utilization.

6.4 Application Summary

Two application case studies are used to demonstrate the applicability and effectiveness of proposed concept. While the experimental lab case study successfully demonstrates how embodied energy storage can be used to increase on-site VRE demand and substitute for battery storage capacity up to a fully autarkical manufacturing line, the second case study demonstrates that enacting energy flexibility can be difficult if process chain layout and parameters are unfavorable. From a methodological point of view, both case studies show different features of the concept and cover different aspects described in Chap. 4. Additional examples for the concept's application can be found in Beier et al. (2015, 2016a, b, 2017).

Computational performance of the simulation prototype strongly varies and is mainly dependent on the process structure complexity and if electricity control is switched on. The amount of different process produce/idle combinations increases exponentially with the number of processes. Further, the electricity control cycle is triggered more frequently as the system state changes more often. For example, one simulation run for the experimental lab simulation over 28 days requires between one and two hours on a workstation-type laptop computer with Intel i7 processor and eight gigabyte RAM. The simulation prototype allows parallel model execution (one model entity per CPU thread), resulting in eight simulation results per one to two hours.

References

Beier, J., Thiede, S., & Herrmann, C. (2015). Increasing energy flexibility of manufacturing systems through flexible compressed air generation. *Procedia CIRP*, *37*, 18–23.

Beier, J., Neef, B., Thiede, S., & Herrmann, C. (2016a). Integrating on-site renewable electricity generation into a manufacturing system with intermittent battery storage from electric vehicles. *Procedia CIRP*, *48*, 483–488.

Beier, J., Thiede, S., & Herrmann, C. (2016b). Integrating variable renewable electricity supply into manufacturing systems. In C. Herrmann & S. Kara (Eds.), *Eco-Factories of the Future (accepted manuscript)*. Berlin: Springer.

Beier, J., Thiede, S., & Herrmann, C. (2017). Energy flexibility of manufacturing systems for variable renewable energy supply integration: Real-time control method and simulation. *Journal of Cleaner Production*, *141*, 648–661.

Bundesnetzagentur (2015). Archivierte Datenmeldungen. Retrieved November 25, 2015, from http://www.bundesnetzagentur.de/cln_1411/DE/Sachgebiete/ElektrizitaetundGas/Unternehmen_Institutionen/ErneuerbareEnergien/Photovoltaik/ArchivDatenMeldgn/ArchivDatenMeldgn_node.html.

Bundesrepublik Deutschland (2014). Erneuerbare-Energien-Gesetz vom 21. Juli 2014 (BGBl. I S. 1066), das durch Artikel 4 des Gesetzes vom 22. Juli 2014 (BGBl. I S. 1218) geändert worden ist.

Heinemann, T. (2016). *Energy and resource efficiency in aluminium die casting*. Switzerland: Springer International Publishing. Cham et al.

Icha, P. (2014). *Entwicklung der spezifischen Kohlendioxid-Emissionen des deutschen Strommix in den Jahren 1990 bis 2013*. Technical report, Umweltbundesamt, Dessau-Roßlau.

Jaspers, J. & Smyrek, D. (2014). Energiekostenbudgetierung an der TU Braunschweig. Retrieved November 25, 2015, from http://www.his-he.de/veranstaltung/dokumentation/Forum/Energie/2014/pdf/13/smyrek/jaspers.pdf.

Lechwerke AG (2015). LEW Grundversorgung Gewerbe. Retrieved November 25, 2015, from https://www1.lew.de/CLP/DOWNLOADS/VERTRAEGE/LEW_A_GVG.PDF.

Stadtwerke Rendsburg (2012). Entgelte für die Netznutzung Strom der Stadtwerke Rendsburg GmbH, gültig ab dem 01.01.2012. Retrieved November 25, 2015, from https://www.stadtwerke-rendsburg.de.

Thiede, S. (2012). *Energy efficiency in manufacturing systems*. Berlin: Springer.

Chapter 7
Summary, Critical Review and Outlook

This chapter closes the book with a summary of main findings and accomplishments, followed by a critical review on limitations. Further, an outlook for potential further improvement and advancement, including future research leads, is given in the last section.

7.1 Summary

Mitigating climate change via switching energy generation from fossil fuel sources to renewable sources constitutes the background for this book. As outlined in Chap. 1, the share of (decentralized) RE electricity generation from non-dispatchable VRE is increasing. Consequently, measures have to be found to integrate VRE into the power system. One option is to align electricity demand to volatile supply, and therefore the top-down research question "How can electricity demand of industry be aligned to (decentralized) variable renewable electricity supply?" is formulated. Based on this research question and additional, industry-centered factors, the bottom-up research question "What are specific methods and tools to enable, implement and evaluate flexible electricity demand of industry, in particular manufacturing systems?" is derived. As a first step, necessary background on production and manufacturing systems, their energy and electricity demand, conventional and renewable electricity supply and corresponding interaction between demand and supply is given in Chap. 2. A focus is set on aligning electricity demand from manufacturing systems to volatile supply. Conventional integration measures such as large- and small-scale energy storage are discussed. Energy flexibility of manufacturing systems (DSM and DR from a utility's perspective) as an approach towards flexible energy/electricity demand from industry is presented. Followed by the current state of the art, research methods and approaches in relation to energy flexibility of manufacturing systems are summarized in Chap. 3. Relevant approaches are identified according to defined selection criteria. These approaches are then classified into single- and multi-process

J. Beier, *Simulation Approach Towards Energy Flexible Manufacturing Systems*, Sustainable Production, Life Cycle Engineering and Management, DOI 10.1007/978-3-319-46639-2_7

systems and their respective planning horizon, including execution-related methods. After providing a detailed review of existing relevant methods, evaluation criteria are defined and applied. The result from this evaluation and classification indicates that approaches which consider multi-process manufacturing systems in combination with multiple planning horizons, including execution, are not available.

Based on identified research demand, a novel energy flexibility planning and improvement method for manufacturing systems is proposed in Chap. 4. The concept is centered around a (longer-term) improvement cycle, which establishes energy flexibility objectives and defines manufacturing system setups which are expected to increase energy flexibility towards set objectives. In order to enable and test hypotheses to achieve objectives, a manufacturing system (energy flexibility) model and control is proposed. The model allows calculation of the manufacturing system evolution and thus determining relevant material and energy flows as well as related performance indicators under different system setups. In addition, a new real-time (execution-related) control concept to enable energy flexibility is described, which can flexibly control manufacturing systems to increase direct demand of decentralized VRE supply while maintaining production throughput. A detailed description of all system elements is given, and complemented by an application cycle to implement proposed concept.

Chapter 5 provides background on the concept's implementation into a simulation prototype. The prototype can be used to test proposed concept, evaluate different scenarios and experiments (system setups) towards energy flexibility objectives and can be easily configured according to a given application case. Two of these application cases are discussed in Chap. 6. Options for achieving an autarkical manufacturing system are determined using an experimental lab factory as example. Another case, an aluminum die casting line, is used to demonstrate additional features of the concept, including stochastic modeling. Finally, this chapter closes with a summary, a critical review and outlook, which are included in the next two sections.

7.2 Critical Review

Although the proposed concept achieves the stated goal of improving energy flexibility of manufacturing systems under decentralized VRE supply, and considers relevant existing technology and advances present research, a critical review is required to understand limitations of proposed concept. Therefore, different topics are reviewed and discussed in the following, separated into **assumption**-, **methodological**- and **applicability/implementation**-related discussion points. The focus of these areas is to highlight limitations, for more detail on what the concept is able to solve and accomplish see Sect. 4.1 "*Concept Objective and Requirements*" as all aspired objectives were achieved.

An integral part of proposed concept is a model centered around manufacturing system energy flexibility. The nature of a model implies that several **assumptions** have been made, which abstract and simplify reality. Within the following, the most important assumption-related topics, which have to be kept in mind during application of the model, are briefly discussed:

- **Process energy demand** modeling accuracy is limited due to abstraction from observable and measured patterns. Proposed state-based modeling is sufficient to resemble real demand within the considered context. Especially in larger systems where realized values from multiple processes are combined and fluctuations are thus evened-out, stochastic influences and deviations from average (state-based) values are less important. Further, installing short-term electricity storage (e.g., capacitors), especially in a VRE-integration context, further smooths profiles. However, especially for processes with multiple different states and thus energy demand, which might be dynamically dependent on each other, more detailed modeling can increase accuracy. For the case of continuous processes, including nonlinear energy demand (as a function of rate) can further enhance fit with observed values. In addition, (very short-term) variations (e.g., overshoot power during start-up) and stochastic variations (e.g., stochastic auxiliary components) are not modeled in detail and could be included with high-resolution and stochastic time series.
- **VRE supply** time series are required for the model. Either recorded time series or physical and mathematical modeling can be applied to obtain desired time series. However, VRE supply profiles underlie a stochastic process and are dependent on (macro-/micro-level) geographic location, time of year and day. As such, applying a suitable (predictive) supply profile is subject to careful consideration of previously mentioned factors. Further, different outcomes with regards to energy flexibility capability and improvement can be expected, depending on which time series is chosen.

On the **methodological** side, the following assumptions and topics need to be considered during application of the concept:

- Dynamic effects on **peripheral systems and infrastructure** are not considered in detail, e.g., dynamics with HVAC (impact on emissions and their regulation) due to energy flexibility actions.
- The proposed **energy control** considers two different energy states of binary processes (adjusted with noncontrollable postproduction state in an application case) and linear adjustment of continuous processes. Adding processes with more controllable energy demand patterns requires adjustment of the control algorithm.
- The defined **bottleneck detection** considers current and past state of the system for bottleneck detection. However, the production program over the simulation time horizon (and thus the target time horizon) needs to be known upfront.
- **Diverting and combining** flows do not create product entities, i.e., although multiple products can be included in the model, their combination or creation (e.g., assembling) is not straightforward possible.
- **Optimization** utilizing proposed concept provides significant challenges. Although a structured target search can be accomplished, finding an optimal solution seems to be unrealistic, as (a) the amount of different input parameters that

can be changed potentially results in system evolution calculation times which are not acceptable, (b) stochastic modeling of several input factors would be required and (c) several target objectives are included, resulting in a multi-objective decision environment and thus a multidimensional target function, which needs to be somehow weighted (if no Pareto-optimum can be found).
- Proposed algorithms for energy flexibility control have an exponentially increasing complexity (and thus computation time) with number of processes and discrete energy states. Therefore, **computational complexity** can become challenging for large/complex systems.

The following **applicability and implementation** related points have been identified:

- The **complexity of the concept** results in additional effort and required resources to adapt the concept to a given application case, which needs to be considered under economic targets (if energy flexibility improvement is subject to economic objectives).
- **Required infrastructure** needs to be installed for machine control and data feedback. Each machine needs to be individually controllable and current status fed back to the control. Further, the central energy control needs to be adapted and included into an (existing) manufacturing execution system.
- In order to achieve energy flexibility via scheduling, **idle capacity/unbalanced** lines are required where production of multiple machines can be shifted to different time periods. This is, for example, the case if a production line is unbalanced (i.e., long idle wait times of processes) or if utilization is low (e.g., due to low demand for products). However, other options such as line re-balancing, reducing operational time of the whole line (reduced total productive time and switch-off), up to fully suspend production can be (economically) preferred options.
- **Conflicting goals** of energy flexibility and lean paradigms, environmental and economic goals might result in reluctance towards implementation. Maximizing throughput, minimizing cost, reducing inventories, and maximizing utilization of equipment are common and central goals of manufacturing companies, which might be less fulfilled due to energy flexibility control actions.
- Although reflected in several operational indicators such as machine and compressor switch-counts, introducing **operational variability and volatility** into existing operations might have adverse effects such as wear-out of machines, degradation of products due to increased storage times or quality impact in general.
- Energy control actions and thus flexibility, variability and tight time-dependencies might impose additional tasks and challenges on **affected staff** (e.g., machine operators, maintenance crews and manufacturing supervisors), for example, resulting in less flexibility regarding task execution time and sequence or break-/shift times. These changes have to be minimized and adequate solutions have to be found together with affected staff.

7.3 Outlook

Several additional leads towards future improvement and advancement of proposed concept have emerged during development of the content presented in previous chapters. They are briefly summarized in the following, clustered by topics on **refinement** of proposed concept, steps towards successful **implementation** and finally leads on **advancement of the concept and potential future research** topics. All discussed topics are evaluated in terms of expected benefit (degree of improvement and advancement) and expected complexity (required resources, time, data and/or methodological challenges).[1]

Based on the assumptions and abstraction made from reality (see also previous section), the following main **refinement** steps are proposed for future work:

R1 **Detailing process energy demand**, which includes multiple (energy) states, time series-based variable energy demand (potentially dependent upon multiple sequential states) and thus complete modeling of energy demand sequences (i.e., depending tasks/task sequence with different energy demand) should be further pursued (cf. assumptions – process energy demand from the critical review section). *(low complexity, medium benefit)*

R2 **Additional peripheral systems**, such as process cooling and HVAC could be included to (a) further detail energy demand modeling of manufacturing systems (which could include mutual influences, e.g., HVAC energy demand as a function of process emissions, which are a function of executed production sequence) and to (b) increase energy flexibility of the system by including additional controllable components (compare to Chap. 3 where approaches towards energy flexibility utilizing HVAC load shifting are named). *(medium complexity, medium benefit)*

R3 **Additional VRE and other supply sources** could be integrated. First, using either physical or mathematical/statistical VRE supply models could be pursuit to reduce reliance on recorded data and test different supply profile characteristics and shapes. Further, adding deployable on-site generation, i.e., diesel backup generation and CHP plants, would refine the concept, but would require an additional control logic for dispatch (see, e.g., Ghadimi et al. 2015 for a control approach to integrate renewable energy sources into factory). *(high complexity, high benefit)*

In order to further pursue **implementation** beyond the developed simulation prototype and example application cases, the following steps could be executed:

I1 **More simulation studies** would allow more generalized conclusions on which parameters influence objectives and other indicators. This might include additional experiments and sensitivity analysis of existing application cases, but also studies involving additional application cases. *(low complexity, medium benefit)*

[1] Note that this basic clustering is done to prioritize and guide future efforts and is based on the subjective assessment of the author.

I2 **More application cases** can further substantiate applicability of proposed concept and would allow to derive generalized conclusions on levers for energy flexibility improvement of manufacturing systems. *(medium complexity, medium benefit)*

I3 After successful additional testing, integrating the control algorithm into a **manufacturing execution system software** requires adaption of developed simulation software. This potentially includes transferring available software code into the programming language of the MES and defining interfaces to MES elements, especially production control. *(high complexity, high benefit)*

I4 Implementing proposed real-time control concept into a **hardware-in-the-loop** simulation would prove its real-time applicability and is expected to reveal challenges towards implementation into a real system which are not discovered in a (pure) software-based simulation. Although this step is not required for successful implementation, it might be able to save effort and resources as challenges are discovered offline, also from the MES software implementation. *(medium complexity, high benefit)*

I5 The implementation into an **industry prototype** and thus into a MES would be the last step towards initial implementation in industry. A suitable manufacturing system needs to be identified, a model representing the system built in a simulation environment (e.g., using the developed software prototype), hypotheses tested and the control algorithm implemented into the MES (if not previously done). The benefit of the approach under realistic (industry-scale) conditions can then be tested and quantified. *(high complexity, high benefit)*

Described refinement and implementation topics are close to and/or based on the current state of proposed concept. The following topics can be addressed for **advancement and future research** related to the concept:

A1 An **advanced control** considering multi-energy states of binary processes and nonlinear energy demand of continuous processes could both increase accuracy of target-state control signals to match a given supply profile (more accurate energy demand prediction) and increase the control range of the central energy control and thus energy flexibility of the system. One lead would be to build a generic approach with a flexible number of different states. Depending on a specific process, a required number of states can then be selected. Further, considering a (nearly) unlimited number of states, nonlinear continuous processes could be modeled using approximation. However, computational complexity (number of control combinations) needs to be kept in mind. *(medium complexity, high benefit)*

A2 Including **operational planning** methods and thus optimization approaches (e.g., from OR, several of them have been discussed in Chap. 3) would complement proposed real-time control method. An optimized planned production program could be used as reference value, short-term (stochastic) disturbances are counteracted by a real-time energy flexibility control. A rolling forecasting of VRE could be applied, and production programs optimized accordingly (similar to, e.g., Schultz et al. 2015). *(high complexity, high benefit)*

A3 Physical and mathematical **VRE time series modeling**, including stochastic modeling, would advance the current approach as more generalized supply time series can be implemented and less reliance on recorded data would be achieved. However, both physical and advanced mathematical models are relatively complex and potentially require significant resources for development. Further, both model types are likely to be stochastic model types, which would require advanced and more complex/time-consuming evaluation and interpretation of results (i.e., determining averages and confidence intervals via Monte Carlo simulation instead of a single value). *(high complexity, medium benefit)*

A4 Extending proposed control concept to include **balancing power applications** would further increase the application range of the concept. The control method and system design could be extended and changed to enable manufacturing companies to participate in reserve markets (primary, secondary, tertiary). However, reserve markets usually have strict (technical) requirements that must be fulfilled, and companies need to commit to supply a reserve capacity if demanded by a system operator or utility, which provides additional challenges that need to be solved. *(high complexity, medium benefit)*

A5 Options to **cluster manufacturing systems and factories** to gain an advanced control range (increased total energy demand control range, more target energy demand states) could be designed to foster energy flexibility in a grid-wide VRE integration context. This approach would bridge the gap between energy flexibility of (a single) manufacturing system and methods known from the supply side (DSM and DR). However, proposed concept would need adaption to multi-factory cases, which requires new methods and control approaches, especially with regards to managing material flow. *(high complexity, high benefit)*

A6 A **theoretical framework** for energy flexibility improvement based on multiple real and synthetic application cases could be developed. The framework would give an indication on what energy storage option (e.g., battery, compressed air, embodied energy in products, thermal storage) yields the best target fulfillment, dependent on manufacturing system setting, energy supply structure and individual objectives. *(low complexity, medium benefit)*

A7 An advanced parallel and dynamic **interaction with a wide range of demand/supply matching** methods and approaches (e.g., fuel switching, grid-wide energy storage such as pumped hydro) can be included to enable a holistic evaluation of all available (technical) options and determine a preferred solution under given objectives. *(high complexity, medium benefit)*

A8 A **dynamic model interacting with the supply side**, i.e., including operational and economic characteristics of an existing power grid and power plant fleet would allow evaluating macro-economic mutual influences and, for example, help to design efficient power supply pricing structures. However, substantial economic modeling would be required. *(high complexity, low benefit)*

A9 Extending the idea of (technical) demand and supply matching methods, a holistic comparison with **all available VRE grid integration measures** could be pursued. A concept for combined comparison of options under multiple (technical, economic, environmental) objectives would be required.

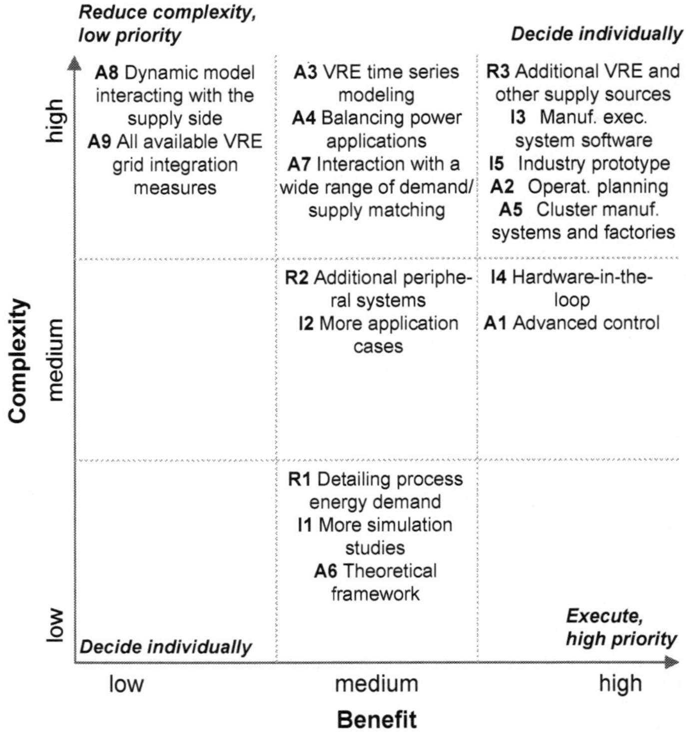

Fig. 7.1 Classification matrix for future advancement and research leads (own illustration)

Combining measures would include all or main methods from central and decentralized energy storage, energy flexibility, DSM and DR (e.g., pricing structures) and reserve capacity deployment (e.g., new fast-response power plants). However, defining comparison methods, building a suitable approach and finally evaluating methods requires substantial effort. *(high complexity, low benefit)*

Figure 7.1 provides a structured summary along the dimensions 'benefit' and 'complexity' of discussed additional leads. This matrix aims at providing some initial guidance for prioritizing future effort to advance the energy flexibility improvement concept presented in this book.

References

Ghadimi, P., Kara, S., & Kornfeld, B. (2015). Renewable energy integration into factories: real-time control of on-site energy systems. *CIRP Annals - Manufacturing Technology*, *64*(1), 443–446.

Schultz, C., Sellmaier, P., & Reinhart, G. (2015). An approach for energy-oriented production control using energy flexibility. *Procedia CIRP*, *29*, 197–202.

Appendix A
Research Evaluation Criteria

Tables A.1, A.2 and A.3 summarize scoring examples per research evaluation criterion to evaluate existing research. A higher score (added quarter to a Harvey Ball) implies that criteria for lower scoring are also fulfilled (cumulative scoring, cf. examples within tables). The given examples are included to provide guidance for scoring. Nonetheless, being a qualitative approach and considering the diversity of reviewed research, not all criterion characteristics can be unambiguously found in reviewed approaches and scoring has been made with regards to best knowledge and belief to allow for a sound evaluation and comparison of results.

Table A.1 Scoring description and examples for *I General* category criteria

Criterion	○	◔	◑	◕	●
I.a General					
I.a.i Assessment and improvement	Energy flexibility not in focus	Qualitative discussion or assessment	Quantitative evaluation	Detailed method to enact or improve energy flexibility	Detailed method to enact or improve energy flexibility, multiple levers
I.a.ii Modularity	No clear modules present	Fixed modular structure	Single changes possible	Multiple simultaneous changes with effort possible	Free and easy combination of elements possible
I.b Planning horizon					
I.b.i Reflection of managerial planning horizons	No explicit consideration of managerial planning process	Single managerial planning horizon considered	Two different planning horizons reflected	Several (more than two) horizons reflected	Several horizons reflected, dependencies considered

(continued)

J. Beier, *Simulation Approach Towards Energy Flexible Manufacturing Systems*, Sustainable Production, Life Cycle Engineering and Management, DOI 10.1007/978-3-319-46639-2

Table A.1 (continued)

Criterion	○	◔	◑	◕	●
I.b.ii Degree of tactical planning	No planning	Consideration of tactical planning	Holistic tactical planning	Consideration with input from other planning stages	Input and output to other planning stages
I.b.iii Degree of operational planning	No operational planning	Consideration of operational planning	Holistic operational planning	Consideration with input from other planning stages	Input and output to other planning stages
I.b.iv Degree of execution	No execution	Consideration of execution	Holistic execution tasks reflected	Consideration with input from other planning stages	Input and output to other planning stages
I.c Temporal dynamics					
I.c.i Changeability of (operational) planning horizon	No planning performed	Fixed planning horizon	Minor adjustments possible	Major adjustments possible	Any time horizon possible
I.c.ii Detail of time resolution	Less than hour intervals	Hour intervals	Hour to minute intervals	Minute intervals	Second intervals
I.c.iii Variability of temporal resolution	No dynamics considered	Fixed timely resolution	Some changes possible	Any timely resolution possible but structural changes required	Any timely resoultion possible
I.c.iv Influence of initial and final model state	No model or no consideration	Influence inherently considered but not discussed nor removed	Influence discussed but not removed	Influence partly removed	No influence of initial and final state

Table A.2 Scoring description and examples for *II System and Structure* category criteria

Criterion	○	◔	◑	◕	●
II.a Electricity system					
II.a.i Changeability of supply parameters and structure	No supply parameters	Fixed supply parameters and structure	Basic changes possible	Multiple scenarios and flexible adjustment possible with major effort	Multiple scenarios and flexible adjustment possible
II.a.ii (Contract) supply terms	No terms provided	Basic terms considered	Main terms provided	Nearly all terms provided	All relevant terms provided
II.a.iii Detail of supply temporal resolution	No dynamic supply	Longer than hour intervals	Hour intervals	Minute intervals	Second intervals
II.a.iv Variability of supply temporal resolution	No dynamic supply	Fixed timely resolution	Some changes possible	Any timely resoultion possible but structural changes required	Any timely resoultion possible
II.a.v Decentralized small-scale energy storage	No dec. small-scale energy storage	Basic consideration of dec. small-scale energy storage	Predefined dec. small-scale energy storage with inherent dynamics	Flexible inclusion of dec. small-scale energy storage	Flexible inclusion of dec. small-scale energy storage, utilized to increase energy flexibility
II.b Manufacturing system					
II.b.i Manufacturing system interaction	Single process, no dependencies	Single process, external dependencies considered	Multiple processes, some dynamics considered	Multiple interacting processes, all relevant internal dynamics considered	Multiple interacting processes, all relevant internal and external dynamics considered
II.b.ii Auxiliary system interaction	No aux. system included	Aux. system considered but not in detail	Aux. system considered in detail, no or minimal dynamics considered	Aux. system modeled in detail, dynamic interaction to manuf. system	Aux. system modeled in detail, dynamic interaction to manuf. system, included in energy flexibility strategy

(continued)

Table A.2 (continued)

Criterion	○	◔	◑	◕	●
II.b.iii Process structure complexity	No structures (single process)	Unidirectional flow structure	Parallel flows, limited number of sub-levels	Any number of sub-levels and divert-ing/converging flows	Diverting, converging and reverse flows possible
II.b.iv Variable process types	Only one process type, basic configuration	One process type, advanced configuration	Both process types, but not combined	Both process types combined possible	Both process types, advanced configuration
II.b.v Electricity demand dynamics	No (individual) electricity demand	Simple on/off electricity demand	Deterministic electricity demand pattern independent of process/system state	Different states and/or time series dependent on system state	Freely adjustable demand patterns and influencing factors

Table A.3 Scoring description and examples for *III Advancement* category criteria

Criterion	○	◔	◑	◕	●
III.a Evaluation					
III.a.i Operational evaluation	No operational evaluation	Qualitative discussion of operational criteria	Basic quantitative operational evaluation, i.e., approximation or single factors	Quantitative evaluation including several factors but not exhaustive	Detailed quantitative evaluation including main influencing factors
III.a.ii Economic evaluation	No economic evaluation	Qualitative discussion of economic criteria	Basic quantitative economic evaluation, i.e., approximation	Quantitative evaluation including several factors but not exhaustive	Detailed quantitative evaluation including main influencing factors
III.a.iii Environmen-tal evaluation	No ecological evaluation	Qualitative discussion of ecological criteria	Basic ecological evaluation, i.e., approximation	Extended ecological evaluation, i.e., quantitative evaluation of main factor(s)	Detailed quantitative evaluation including main influencing factors
III.b Improvement					
III.b.i Adjustability of input parameters and structure	No improvement targeted	Single or very limited number of input parameters	Main input parameters or limited parameters but basic structure adjustable	Main input parameters and basic structure adjustable	Multiple input parameters and freely adjustable structure

(continued)

Table A.3 (continued)

Criterion	○	◔	◑	◕	●
III.b.ii Flexibility of objective definition	No improvement targeted	Single, predefined objective	Single, but flexibly definable objective	Multiple, predefined objectives	Multiple, flexibly definable objectives
III.b.iii Input uncertainty and sensitivities	No improvement targeted or single parameter set	Simple sensitivity analysis performed or possible	Systematic sensitivity analysis (multi-factor sensitivities) performed or possible	Randomness (probability distr.) in single input parameter category evaluated	Randomness (probability distr.) in (multiple) input parameter categories considered
III.c Applicability					
III.c.i Visualization of results	No visualization	Reporting of numbers	Reporting of simple charts	Holistic graphical reporting of indicators	Holistic and dynamic graphical reporting of indicators
III.c.ii Implemented solution	No implemented solution	Solution for example case inferred	Implemented solution, but centered around specific method or example	Modular and universally applicable solution	Modular and universally applicable prototype which can be rapidly applied
III.c.iii Easiness of configuration	Not configurable	Expert knowledge required	Teaching by expert or manual required	Some knowledge and time required	Intuitively and rapidly configurable
III.c.iv Easiness of parameterization	Not configurable	Expert knowledge required	Teaching by expert or manual required	Some knowledge and time required	Intuitively and rapidly parameterized, automation available
III.c.v Intuitively and safe to use	No target user	Expert knowledge required, method applicable with reasonable effort	Application and underlying logic intuitively and fast understandable	Application and underlying logic intuitively and fast understandable, underlying assumptions clear and safe to apply	Application and underlying logic intuitively understandable, integrity checks